A Generalization of Riemann Mappings and Geometric Structures on a Space of Domains in $\mathbf{C}^n$

Recent Titles in This Series

(Continued in the back of this publication)

MEMOIRS
of the
American Mathematical Society

Number 472

A Generalization of Riemann Mappings and Geometric Structures on a Space of Domains in $\mathbf{C}^n$

Stephen Semmes

July 1992 • Volume 98 • Number 472 (third of 4 numbers) • ISSN 0065-9266

American Mathematical Society
Providence, Rhode Island

1991 *Mathematics Subject Classification.*
Primary 32H99; Secondary 32G99, 32F07.

Library of Congress Cataloging-in-Publication Data

Semmes, Stephen, 1962–
A generalization of Riemann mappings and geometric structures on a space of domains in $\mathbf{C}^n$/Stephen Semmes.
p. cm. – (Memoirs of the American Mathematical Society, ISSN 0065-9266; no. 472)
On t.p. n is superscript.
"July 1991, volume 98, number 472 (third of 4 numbers)."
Includes bibliographical references.
ISBN 0-8218-2532-1
1. Riemann surfaces. 2. Conformal mapping. I. Title. II. Series.
QA3.A57 no. 472
[QA333]
500 s–dc20 92-12014
[515′.223] CIP

Memoirs of the American Mathematical Society

This journal is devoted entirely to research in pure and applied mathematics.

Subscription information. The 1992 subscription begins with Number 459 and consists of six mailings, each containing one or more numbers. Subscription prices for 1992 are \$292 list, \$234 institutional member. A late charge of 10% of the subscription price will be imposed on orders received from nonmembers after January 1 of the subscription year. Subscribers outside the United States and India must pay a postage surcharge of \$25; subscribers in India must pay a postage surcharge of \$43. Expedited delivery to destinations in North America \$30; elsewhere \$82. Each number may be ordered separately; *please specify number* when ordering an individual number. For prices and titles of recently released numbers, see the New Publications sections of the *Notices of the American Mathematical Society.*

Back number information. For back issues see the *AMS Catalogue of Publications.*

Subscriptions and orders should be addressed to the American Mathematical Society, P. O. Box 1571, Annex Station, Providence, RI 02901-1571. *All orders must be accompanied by payment.* Other correspondence should be addressed to Box 6248, Providence, RI 02940-6248.

Memoirs of the American Mathematical Society is published bimonthly (each volume consisting usually of more than one number) by the American Mathematical Society at 201 Charles Street, Providence, RI 02904-2213. Second-class postage paid at Providence, Rhode Island. Postmaster: Send address changes to Memoirs, American Mathematical Society, P. O. Box 6248, Providence, RI 02940-6248.

Printed in the United States of America.
This volume was printed directly from author-prepared copy.
The paper used in this book is acid-free and falls within the guidelines
established to ensure permanence and durability. ♾
10 9 8 7 6 5 4 3 2 1 97 96 95 94 93 92

CONTENTS

ABSTRACT

The idea behind this generalization of Riemann mappings is to try to have a class of mappings that preserve complex structures to as large an extent as possible, subject to the constraint that it be flexible enough for there to be a decent existence theory. The notion of Riemann mappings considered here is closely related to one introduced by Lempert a decade ago. An important difference between the two approaches is that the notion of Riemann mapping given here is essentially characterized by a differential equation. On the other hand, it is shown herein that there is a way to pass back and forth between these two approaches, which is important because it makes Lempert's analytical results available for the present purposes.

One of the main reasons for looking for a good theory of Riemann mappings is that it would be helpful for having something for domains in $\mathbf{C}^n$ which is analogous to the universal Teichmüller space for $n = 1$. This works out better than you might expect. The space of Riemann mappings admits, at least in a weak sense, a complex structure and a degree of local homogeneity that correspond very naturally to results in the classical $n = 1$ case. There is also a natural Riemannian metric.

Symplectic structures are one of the key ingredients in this story. A simple way in which they arise is the old idea that a symplectomorphism on a cotangent bundle can be viewed as a generalization of a diffeomorphism on the underlying manifold. This is relevant in particular for the local homogeneity on the space of Riemann mappings that was referred to above.

Key words and phrases. Riemann mappings, complex Monge-Ampère, symplectic structures, complex structures, spaces of domains.

1. INTRODUCTION

Let B_n denote the unit ball in $\mathbf{C}^n$, $n > 1$. We shall call a mapping $\rho : B_n \to \mathbf{C}^n$ a Riemann mapping if it satisfies the four conditions (1.1)–(1.4) described below. The idea is to allow more than just biholomorphisms, in order to get some kind of general existence results, but to still respect the complex structure in some way. The first two conditions are (negotiable) regularity requirements; (1.3) and (1.4) are the key conditions.

(1.1) *ρ is a homeomorphism of B_n onto its image, and $\rho(0) = 0$.*

(1.2) *ρ is C^1 on $B_n \backslash \{0\}$, its differential $d\rho$ is invertible at all points in $B_n \backslash \{0\}$, and ρ is bilipschitz on $\frac{1}{2}B_n$.*

Recall that a mapping f is bilipschitz if there is a constant $C > 0$ so that

$$C^{-1}|a - b| \le |f(a) - f(b)| \le C|a - b|$$

for all a, b in the domain of f. It is not hard to show that (1.1) and (1.2) imply that ρ is bilipschitz on rB_n for all $r < 1$. It is important that we do not require ρ to be C^1 at the origin.

(1.3) *For each $z \in \partial B_n$, $\lambda \mapsto \rho(\lambda z)$ is a holomorphic map of the unit disk Δ into $\mathbf{C}^n$.*

Before stating the last condition it is helpful to introduce some notation (which will be used throughout). For $z \in \mathbf{C}^n$, $z \neq 0$, set

$$S_z^1 = \{v \in \mathbf{C}^n : v = \lambda z \text{ for some } \lambda \in \mathbf{C}\},$$
$$S_z^2 = \{v \in \mathbf{C}^n : \Sigma v_j \bar{z}_j = 0\}.$$

Thus $T_z\mathbf{C}^n \cong \mathbf{C}^n = S_z^1 + S_z^2$, and (1.3) is the same as saying that $d\rho_z$ is complex-linear on S_z^1.

(1.4) *For each $z \in B_n \setminus \{0\}$, $d\rho_z$ maps S_z^2 to a complex subspace of $\mathbf{C}^n$.*

Received by the editors October 3, 1990. Received in revised form July 3, 1991

The author is partially supported by the NSF, the Alfred P. Sloan Foundation, and the Marian and Speros Martel Foundation.

This last condition has a nice reformulation if you assume also that ρ extends to a homeomorphism of $\overline{B}_n$ onto its image which is a C^1 diffeomorphism on $\overline{B}_n \setminus \{0\}$. In that case (1.4) holds for all $z \in \partial B_n$ if and only if ρ restricts to a contact mapping between ∂B_n and $\rho(\partial B_n)$, where these two hypersurfaces are given their usual contact structures (induced by the complex structure on $\mathbf{C}^n$). Conversely, the requirement that $\rho : \partial B_n \to \rho(\partial B_n)$ be a contact mapping implies (1.4) in the presence of (1.1)–(1.3); this is not hard to derive from the equivalence of (1.4) with condition (1.5), stated below, using the fact that a holomorphic function on $\Delta \setminus \{0\}$ cannot vanish on $\partial\Delta$ without vanishing identically.

This notion of a Riemann mapping turns out to be closely related to one introduced by L. Lempert in [L1]. In fact it is possible to pass back and forth between the two notions of a Riemann mapping, in a certain sense that will be made precise later. This will permit us to use the work of Lempert to get existence results. One of the primary differences between this approach to Riemann mappings and Lempert's is that the main conditions here — (1.3) and (1.4) — can be rewritten in terms of first order partial differential equations.

We shall see that it follows from [L1, **3**] that every smooth, strongly convex domain D in $\mathbf{C}^n$ which contains the origin arises as the image of a Riemann mapping. We shall also see that Riemann mappings always have a close relationship with Green's functions for the complex Monge-Ampère operator and extremal holomorphic maps of the disk into the image domain, as they do when they are obtained from Lempert's work. We shall derive other properties and characterizations of Riemann mappings, and prove in particular that their image is always pseudoconvex.

The regularity assumptions in (1.2) were chosen to balance considerations of generality against convenience. In order to obtain general existence results, perhaps through a variational principle, it would probably be necessary to allow mappings that have substantially less regularity, and to modify (1.4) accordingly. We shall not address this issue here, but we shall present equivalent characterizations of Riemann mappings that may be better suited to dealing with the problem of existence of weak solutions. As far as that goes, the above reformulation of (1.4), in which ρ is required to define a contact mapping of ∂B_n onto $\rho(\partial B_n)$, may be amenable to a variational approach.

The uniqueness issue is easier to resolve than existence. We shall show that if two Riemann mappings have the same image, then they differ by a map of the ball to itself which lies in a certain group. This map must in particular commute with dilations by complex numbers, and so there can be at most one Riemann mapping whose image and first-order behavior at the origin are prescribed, just as in the $n = 1$ case.

There are other properties which these Riemann mappings have that are analogous to properties of conformal mappings in $\mathbf{C}$. For example, using [L4] we shall show that if $\rho : B_n \to \mathbf{C}^n$ satisfies (1.1)–(1.4) and extends to a bilipschitz map of $\overline{B}_n$ into $\mathbf{C}^n$ that is real-analytic on $\overline{B}_n \setminus \{0\}$, then ρ induces a nonlinear transformation on a set of functions so that solutions of the homogeneous complex Monge-Ampère equation (hereafter referred to as HCMA) are taken to solutions of HCMA. These transformations should be viewed as generalizations of $f \mapsto f \circ \rho$, and they reduce to such a composition when ρ is holomorphic.

Another general area of similarity between the $n = 1$ and the $n > 1$ cases concerns the space of Riemann mappings. For example, when $n = 1$ the space of conformal mappings $f : \Delta \to \mathbf{C}$ with $f(0) = 0$ and suitable regularity conditions imposed on $\partial\Delta$ can be realized as an open subset of a complex Banach space, which provides a complex structure on this space of Riemann mappings. When $n > 1$ we can realize the space of mappings $\rho : B_n \to \mathbf{C}^n$ that satisfy (1.1)–(1.3), plus suitable regularity conditions on ∂B_n, as an open subset of a complex Banach space of mappings. Although (1.4) is neither an open nor a linear condition, it can be expressed as a holomorphic constraint. Let ϕ denote the complex n-form $dz_1 \wedge \cdots \wedge dz_n$ on $\mathbf{C}^n$, and let $\delta\rho(\phi)$ denote the pull-back of ϕ using ρ. Then (1.4) is equivalent to

(1.5) *the restriction of* $\delta\rho(\phi)\big|_z$ *to* S_z^2 *vanishes for each* $z \in B_n \setminus \{0\}$.

[Indeed, if L is a real linear subspace of $\mathbf{C}^n$ with real dimension $2n - 2$, then L is complex if and only if ϕ restricts to zero on L.] Because

$$\rho \mapsto \delta\rho(\phi)$$

is a holomorphic homogeneous polynomial mapping of degree n in ρ, this says that the space of Riemann mappings is an (infinite-dimensional) complex variety, at least if we interpret "variety" liberally, as we shall throughout.

Unfortunately, it is not clear how to give this space the structure of a complex Banach (or Frechét) manifold. In Section 8 we shall encounter another description of the space of Riemann mappings as the zero set of a holomorphic polynomial mapping (which will in fact be quadratic), and although this description appears to be better-suited to implicit function theorem techniques, it is still not clear how to get a nice manifold structure in, say, the C^∞ category. One probably could make implicit function theorem techniques work in the real-analytic category using the observations of Section 8, but this would not be so nice, because the real-analytic category is not a good place to be when you want to show that an infinite-dimensional space is a manifold. In the real-analytic category there are other, simpler, methods for addressing this issue. There is an exponentiation process (discussed in Section 8) that can be used to generate plenty of holomorphic families of Riemann mappings, for instance. Using the method of generating functions we can also, in some sense, parameterize the space of real-analytic Riemann mappings that are close to a given one, in a way that is compatible with the complex structures. This will also be explained in Section 8.

In practice we shall often be able to get around the lack of a nice manifold structure, even in the C^∞ category, in reasonable ways. For instance, if a given function maps into a space of Riemann mappings, then we shall consider it to be holomorphic if it is holomorphic as a map into the ambient complex vector space, while a function defined on a subset of a space of Riemann mappings will be considered holomorphic it if admits a holomorphic extension to an open subset of the ambient space. These conventions serve well in the situations that arise here.

When $n = 1$ the space of Riemann mappings admits a rich class of natural local holomorphic transformations, given by composition by conformal maps on the left. That is, if g is a conformal map from some domain in $\mathbf{C}$ to another, and $g(0) = 0$, then $f \mapsto g \circ f$ defines a transformation on a subset of the space of Riemann mappings into another such subset. There is a version of this when $n > 1$ which is more complicated (and closely related to [L4]), and we shall study this action in some detail.

We are also going to look at the space of domains in $\mathbf{C}^n$ that arise as images of Riemann mappings, or, rather, the mappings that have some extra smoothness. Some interesting structures on this space will be inherited from the space of Riemann mappings. For example, the space of domains will have some interesting structure that derives from the realization of the space of Riemann mappings as a complex variety. We shall also put a Riemannian metric on the space of domains defined in terms of an L^2 norm of variations of the Green's function for the complex Monge-Ampère operator. This metric has a number of nice properties, including the fact that it is preserved by the action on domains induced by the action on the space of Riemann mappings mentioned above.

The absence of a manifold structure on the space of Riemann mappings will be more troublesome when we study this Riemannian metric on the space of their images. For this reason we shall often restrict ourselves in this context to smooth, strongly convex domains, so that Lempert's work can be used to justify the formal calculations.

Completely circled domains will play a special and prominant role throughout this paper. [A domain D is called completely circled if $\lambda D = \{\lambda z : z \in D\}$ is contained in D for all $\lambda \in \overline{\Delta}$.] These domains arise as both a subset of the space of domains, and as an image, via the Kobayashi indicatrix. (See Section 3.) The Riemannian metric on the space of completely circled domains that we consider here was encountered previously in [S], and it seems to be very natural. (See also [CS].)

We discuss the properties of Riemann mappings in Sections 2 through 8, and we study spaces of Riemann mappings and domains in Sections 9 through 21. A list of some notations and conventions employed in this paper is included just before the references.

The author would like to thank John Bland and Tom Duchamp for helpful discussions and suggestions.

2. RIEMANN MAPPINGS, GREEN'S FUNCTIONS, AND EXTREMAL DISKS

Fix $\rho : B_n \to \mathbf{C}^n$. Define F_0, u_0 on B_n by

$$F_0(z) = |z|^2 = \Sigma |z_j|^2, \quad u_0(z) = \log |z|. \tag{2.1}$$

Set $D = \rho(B_n)$, and define F, u on D by $F = F_0 \circ \rho^{-1}$, $u = u_0 \circ \rho^{-1}$.

THEOREM 2.2. *Assume that ρ satisfies (1.1) and (1.2). Then the following are equivalent:*

(a) *ρ satisfies (1.3) and (1.4);*
(b) *$\delta\rho(\partial F) = \partial F_0$ on $B_n \setminus \{0\}$;*
(c) *$\delta(\rho^{-1})(\partial F_0)$ is a $(1,0)$ form on $D \setminus \{0\}$.*

THEOREM 2.3. *Suppose that ρ satisfies (1.1)–(1.4). Then:*

(a) *D is pseudoconvex, and if ρ extends to a bilipschitz mapping on $\overline{B}_n$ that is smooth on ∂B_n, then D is strongly pseudoconvex;*
(b) *u is plurisubharmonic on D;*
(c) *$\partial\overline{\partial}u$ is continuous on $D \setminus \{0\}$, and it satisfies $(\partial\overline{\partial}u)^n = 0$, $(\partial\overline{\partial}u)^{n-1} \neq 0$ there;*
(d) *$u(z) - \log|z|$ is bounded, and $u(z) \to 0$ as $z \to \partial D$;*
(e) *u is the Green's function for the complex Monge-Ampère operator on D with pole at the origin.*

In (c) we mean that $\partial\overline{\partial}u$, viewed as a distribution (or, more precisely, a current), is actually given by a two-form with continuous coefficients. For (e) recall that the Green's function for the complex Monge-Ampère operator with pole at the origin (hereafter referred to simply as the Green's function) on D is given by

(2.4) $\sup\{v : v$ is a plurisubharmonic function on D, $v \leq 0$ on D, and $v(z) \leq \log|z| + C$ for $z \in D$ and some constant $C < \infty$ that depends on $v\}$.

In particular, (e) implies that if ρ satisfies (1.1)–(1.4), then u and F depend only on D, and not on the particular choice of ρ.

THEOREM 2.5. *Suppose that ρ satisfies (1.1)–(1.4). Suppose also that $f : \Delta \to D$ is given by $f(\lambda) = \rho(\lambda v)$ for some $v \in \partial B_n$. Then f is extremal in the following sense. If $g : \Delta \to D$ is holomorphic and satisfies $g(0) = 0$, $g'(0) = af'(0)$ for some $a \in \mathbf{C}$, then $|a| \leq 1$. If $|a| = 1$, then $g(\lambda) = f(a\lambda)$.*

Theorems 2.2 and 2.3 suggest variations of the definition of Riemann mappings that might be better suited to weaker regularity assumptions. For instance, you could demand that ρ be sufficiently well-behaved so that $\delta(\rho^{-1})(\partial F_0)$ makes sense as a current and that it be a $(1,0)$ form. Alternatively, if D is sufficiently nice that it has a Green's function u, then you could require that $\delta\rho(\partial F) = \partial F_0$ in a distributional sense, where $F = \exp(2u)$. [Conditions on a domain that ensure that the Green's function (2.4) is at least moderately well-behaved are given in [D], [K]; it is sufficient to assume that D is smooth and strongly pseudoconvex.]

The rest of this section is devoted to the proofs of these three theorems.

Let's start with Theorem 2.2. Clearly (b) implies (c). The converse is easy too; because $F = F_0 \circ \rho^{-1}$, we have $dF = \delta(\rho^{-1})(dF_0)$, and so we must have $\partial F = \delta(\rho^{-1})(\partial F_0)$ if $\delta(\rho^{-1})(\partial F_0)$ is a $(1,0)$ form, since ∂F is the unique $(1,0)$ form whose real part is $\frac{1}{2}dF$.

It is also not hard to show that (a) implies (c), using the fact that S_z^2 is the kernel of $\partial F_0 \big|_z$. It remains to show that (b) implies (a).

Suppose that $\delta\rho(\partial F) = \partial F_0$. This implies that for each $z \in B_n \setminus \{0\}$, $d\rho_z$ maps S_z^2 to the kernel of $\partial F \big|_{\rho(z)}$, which implies (1.4). To check (1.3) requires more work.

Given $w \in D \setminus \{0\}$, let q_w denote the obvious quotient map from $T_w\mathbf{C}^n \cong \mathbf{C}^n$ onto $T_w\mathbf{C}^n / T_w^2$, where T^2 is the subbundle of $T\mathbf{C}^n$ on $D \setminus \{0\}$ which is the image of S^2 on $B_n \setminus \{0\}$ under $d\rho$. It is not difficult to show that $q_{\rho(z)} \circ d\rho_z$ is complex linear on S_z^1 for each $z \in B_n \setminus \{0\}$, using $\delta\rho(\partial F) = \partial F_0$. If we can show that $d\rho_z(S_z^1)$ is a complex subspace of $\mathbf{C}^n$, then it will follow that $d\rho_z$ is complex-linear on S_z^1, so that (1.3) holds. To do this we need an auxiliary fact that will also be used later.

LEMMA 2.6. *Suppose that ρ satisfies (1.1), (1.2), and $\delta\rho(\partial F) = \partial F_0$. Set $\omega_0 = \frac{1}{2i}\partial\overline{\partial} F_0$ and $\omega = \frac{1}{2i}\partial\overline{\partial} F$, where ω is a priori only a current. Then in fact ω has continuous coefficients on $D \setminus \{0\}$, and*

$$\omega = \delta(\rho^{-1})(\omega_0). \tag{2.7}$$

This is basically trivial. Formally we have

$$\omega = \frac{i}{2} d\partial F = \frac{i}{2} d(\delta(\rho^{-1})(\partial F_0)) = \delta(\rho^{-1})\left(\frac{i}{2} d\partial F_0\right) = \delta(\rho^{-1})(\omega_0).$$

To make this rigorous we use an approximation argument. Let $\psi_j : D \setminus \{0\} \to \mathbf{C}^n$ be a sequence of smooth mappings that converges to ρ^{-1} in the C^1 topology on every compact subset of $D \setminus \{0\}$. Then we certainly have that

$$\frac{i}{2} d(\delta\psi_j(\partial F_0)) = \delta\psi_j(\omega_0). \tag{2.8}$$

Clearly $\delta\psi_j(\omega_0) \to \delta(\rho^{-1})(\omega_0)$ and $\delta\psi_j(\partial F_0) \to \delta(\rho^{-1})(\partial F_0)$ uniformly on compact subsets of $D \setminus \{0\}$. The lemma now follows by taking the limit as $j \to \infty$ of (2.8) in the sense of currents.

Let us use (2.7) to prove that $d\rho_z(S_z^1)$ is a complex subspace for all $z \in B_n \setminus \{0\}$. By definitions,

$$S_z^1 = \{v \in \mathbf{C}^n : (\omega_0 \big|_z)(v, v') = 0 \qquad \text{for all } v' \in S_z^2\},$$

and hence

$$d\rho_z(S_z^1) = \{v \in \mathbf{C}^n : \left(\omega \big|_{\rho(z)}\right)(v, v') = 0 \qquad \text{for all } v' \in T^2_{\rho(z)}\}.$$

This is a complex subspace because $T^2_{\rho(z)}$ is, and because ω is a $(1,1)$ form. This finishes the proof of Theorem 2.2.

The main step in the proof of Theorem 2.3 is the following.

LEMMA 2.9. *Suppose that ρ satisfies (1.1)–(1.4). Then for each $z \in D \setminus \{0\}$, $\frac{\partial^2}{\partial z_j \partial \bar{z}_k} F(z)$ is a positive-definite matrix.*

Notice that $\frac{\partial^2}{\partial z_j \partial \bar{z}_k} F(z)$ is continuous in z on $D \setminus \{0\}$, because of Lemma 2.6.

Let $A(z)$ denote the self-adjoint complex-linear transformation on $\mathbf{C}^n$ associated to this matrix. For each z in $D \setminus \{0\}$ $A(z)$ has trivial kernel, because of (2.7) and the fact that ω_0 is nondegenerate. Hence $A(z)$ must have a negative eigenvalue for all z if it has one for some z, since $A(z)$ is continuous in z. Let us assume that this is true, and get a contradiction using the minimum principle. We have to be a little careful, because of the singularity at the origin.

For each $\epsilon > 0$ set

$$f_\epsilon(z) = F(z) - \epsilon \log|z|$$

on $D \setminus \{0\}$, and let $A_\epsilon(z)$ be the linear transformation associated to $\frac{\partial^2}{\partial z_j \partial \bar{z}_k} f_\epsilon(z)$. Because $\log|z|$ is plurisubharmonic $A_\epsilon(z)$ has at least one negative eigenvalue for each $\epsilon > 0$ and $z \in D \setminus \{0\}$. This implies that

$$\inf_{z \in U} f_\epsilon(z) = \inf_{z \in \partial U} f_\epsilon(z) \tag{2.10}$$

for each open set U with $\overline{U} \subseteq D \setminus \{0\}$. This is actually a little tricky; we cannot use the standard minimum principle directly, because f_ϵ may not be C^2, even though $\frac{\partial^2}{\partial z_j \partial \bar{z}_k} f_\epsilon$ is continuous. However, this difficulty is easily overcome by a standard approximation argument.

It is easy to see that (2.10) is impossible if ϵ is very small and U is almost all of $D \setminus \{0\}$, because $F = 1$ on ∂D, $F < 1$ on D, and $f_\epsilon(0) = \infty$ for all $\epsilon > 0$. This contradiction establishes Lemma 2.9.

Let us now prove Theorem 2.3. Suppose that ρ satisfies (1.1)–(1.4). Then F is plurisubharmonic on D, because of Lemma 2.9, and we have $F < 1$ on D, $F(z) \to 1$ as $z \to \partial D$, which implies that D is pseudoconvex.

If ρ is bilipschitz on B_n, then the eigenvalues of $A(z)$ can be bounded away from 0 on $D \setminus \{0\}$, because of (2.7) and the nondegeneracy of ω_0. If ρ is also smooth on a neighborhood of ∂B_n, then it follows easily that D is strongly pseudoconvex. [Extend F smoothly to a neighborhood of ∂D, and use $F - 1$ as a defining function.] This proves (a).

Let us check (c) before (b). Observe that

$$\partial\overline{\partial}u = \partial\overline{\partial}(\log F) = \partial(F^{-1}\overline{\partial}F) = F^{-1}\partial\overline{\partial}F - F^{-2}\partial F \wedge \overline{\partial}F. \tag{2.11}$$

These equalities hold in the sense of currents on $D \setminus \{0\}$. They can be justified rigorously using an approximation argument and the fact (from Lemma 2.6) that $\partial\overline{\partial}F$ is continuous. This implies that $\partial\overline{\partial}u$ is continuous, and that

$$\partial\overline{\partial}u = \delta(\rho^{-1})(\partial\overline{\partial}u_0), \tag{2.12}$$

which could also have been proved exactly as in Lemma 2.6. Because $(\partial\overline{\partial}u_0)^n = 0$, $(\partial\overline{\partial}u_0)^{n-1} \neq 0$ on $B_n \setminus \{0\}$, the corresponding results hold for u, which gives (c).

To prove (b) it suffices to show that $\frac{\partial^2}{\partial z_j \partial \overline{z}_k} u(z)$ is a nonnegative matrix for all $z \in D \setminus \{0\}$. Because of (2.11) and Lemma 2.9, it can have at most one non-positive eigenvalue. On the other hand, (c) says that it must have a zero eigenvalue, whence the desired nonnegativity.

It remains to verify (e), since (d) is immediate from the definitions. Clearly u is a competitor for the supremum in (2.4), by (b) and (d), and so u is not greater than (2.4). We need to show that $u \geq v$ for all v as in (2.4).

Fix v. Given $z \in \partial B_n$, define a real-valued function h on Δ by

$$h(\lambda) = v(\rho(\lambda z)).$$

Then $h \leq 0$ on Δ, h is subharmonic, and $h(\lambda) \leq \log|\lambda| + C$, where C depends on v and ρ but not λ. This implies that $h(\lambda) \leq \log|\lambda|$ on Δ. [For each $\epsilon > 0$ set $h_\epsilon(\lambda) = h(\lambda) - (1-\epsilon)\log|\lambda|$. Then h_ϵ is subharmonic on $\Delta \setminus \{0\}$ and $h_\epsilon(0) = -\infty$, and so h_ϵ is subharmonic on Δ. The maximum principle implies that $h_\epsilon \leq 0$ for each $\epsilon > 0$, and hence $h(\lambda) \leq \log|\lambda|$.]

On the other hand we have $u(\rho(\lambda z)) = \log|\lambda|$. Because z, λ are arbitrary we obtain $u \geq v$. This completes the proof of Theorem 2.3.

Let us now turn to the proof of Theorem 2.4. Let v, f, g, and a be as in the statement of Theorem 2.4, and assume that ρ satisfies (1.1)–(1.4). We may as well require that $g'(0) \neq 0$.

Consider the function

$$s(\lambda) = u(g(\lambda)) - \log|\lambda|$$

defined on $\Delta \setminus \{0\}$. By Theorem 2.3(b) $s(\lambda)$ is subharmonic on $\Delta \setminus \{0\}$. It is bounded in a neighborhood of 0, because of (1.2) and the definition of u, and so it has a subharmonic extension to Δ. [One way to check this uses the fact $s(\lambda) + \epsilon \log|\lambda|$ is subharmonic on Δ for all $\epsilon > 0$.] Hence

$$\limsup_{\lambda \to 0} s(\lambda) \leq \limsup_{\lambda \to \partial\Delta} s(\lambda) = 0.$$

By definition of u we have that

$$s(\lambda) = \log\{|\lambda|^{-1}|\rho^{-1}(g(\lambda))|\}.$$

Also, $g(\lambda) - \rho(\lambda av) = 0(|\lambda|^2)$ as $\lambda \to 0$, and so $\rho^{-1}(g(\lambda)) - \lambda av = 0(|\lambda|^2)$. Hence $0 \geq \limsup_{\lambda \to 0} s(\lambda) = |a|$, and so $|a| \leq 1$.

Suppose that $|a| = 1$, so that $s(\lambda) \equiv 0$, by the maximum principle. The first step in proving that $g(\lambda) = f(a\lambda)$ is to show that

$$g'(\lambda) \in T^1_{g(\lambda)} \qquad \text{for each } \lambda \neq 0, \tag{2.13}$$

where T^1_z is the subbundle of the tangent space of $D \setminus \{0\}$ which is the image of S^1 under $d\rho$.

Because $s(\lambda) \equiv 0$, we have that $u(g(\lambda)) = \log|\lambda|$ is harmonic on $\Delta \setminus \{0\}$. We'll derive (2.13) from this and the fact that $\left(\frac{\partial^2}{\partial z_j \partial \bar{z}_k} u(z)\right)_{j,k}$ has only one zero eigenvalue (and the rest positive) for each z, but we have to do this slowly because u is not necessarily C^2.

Fix $\lambda_0 \in \Delta \setminus \{0\}$. Thus $u(g(\lambda_0)) = \log|\lambda_0| > -\infty$, and so $g(\lambda_0) \neq 0$. Let u_l be a sequence of C^∞ functions defined on a neighborhood of $g(\lambda_0)$ such that $u_l \to u$ and $\partial\bar{\partial} u_l \to \partial\bar{\partial} u$ uniformly. For λ near λ_0 we have

$$\frac{\partial^2}{\partial \lambda \partial \bar{\lambda}}(u_l(g(\lambda))) = \sum_{j,k} \left(\frac{\partial^2}{\partial z_j \partial \bar{z}_k} u_l\right)(g(\lambda)) g_j'(\lambda) \overline{g_k'(\lambda)}.$$

As $l \to \infty$ the right side tends to

$$\sum_{j,k} \left(\frac{\partial^2}{\partial z_j \partial \bar{z}_k} u\right)(g(\lambda)) g_j'(\lambda) \overline{g_k'(\lambda)} \tag{2.14}$$

uniformly, while the left side tends to 0 in the sense of distributions in λ, because $u_l(g(\lambda))$ tends uniformly to $u(g(\lambda)) = \log|\lambda|$. Thus (2.14) vanishes on a neighborhood of λ_0, and hence on all of $\Delta \setminus \{0\}$, since λ_0 was arbitrary.

Let $L(z)$ denote the linear map associated to the matrix $\frac{\partial^2}{\partial z_j \partial \bar{z}_k} u(z)$, $z \in D \setminus \{0\}$. We saw in the proof of Theorem 2.3 that $L(z)$ is nonnegative for each z, and so $g'(\lambda)$ lies in the kernel of $L(g(\lambda))$, because (2.14) vanishes.

The kernel of $L(z)$ consists of the set of vectors $w \in T_z\mathbf{C}^n \cong \mathbf{C}^n$ such that

$$i(w)(\partial\bar{\partial} u \,|_z) = 0,$$

where $i(w)$ denotes the interior product by w. This is also the same as T^1_z. To see this, let $z' = \rho^{-1}(z)$, fix $w \in T_z\mathbf{C}^n$, and let $w' \in T_{z'}\mathbf{C}^n$ be such that $d\rho_{z'}(w') = w$. Then $i(w)(\partial\bar{\partial} u \,|_z) = 0$ iff $i(w')(\partial\bar{\partial} u_0 \,|_{z'}) = 0$, because of (2.12). This last occurs iff $w' \in S^1_z$, which is equivalent to $w \in T^1_z$, From here (2.13) follows easily.

Because T^1_z is always a complex one-dimensional subspace of $T_z\mathbf{C}^n$, (2.13) implies that the tangent space of $g(\Delta)$ at $g(\lambda)$ is $T^1_{g(\lambda)}$ for all $\lambda \neq 0$. Thus if

$h : \Delta \to B_n$ is defined by $h = \rho^{-1} \circ g$, then h is holomorphic away from the origin, because $d\rho$ is a complex-linear map of S^1 to T^1. Because h is bounded it is also holomorphic at the origin.

Let's determine $h'(0)$. Because $g'(0) = af'(0)$, we have that $g(\lambda) - f(a\lambda) = 0(|\lambda|^2)$ as $\lambda \to 0$, and so $g(\lambda) - \rho(\lambda a v) = 0(|\lambda|^2)$ by definition of f. Hence $h(\lambda) - \lambda a v = 0(|\lambda|^2)$ also, and so $h'(0) = av$.

Now we are essentially finished. Define $\tilde{h} : \Delta \to \mathbf{C}^n$ by $\tilde{h}(\lambda) = \lambda^{-1}h(\lambda)$ when $\lambda \neq 0$, $\tilde{h}(0) = h'(0)$. Then $\tilde{h}$ is holomorphic and $\lim_{\lambda \to \partial\Delta} |\tilde{h}(\lambda)| \leq 1$, since $h(\Delta) \subseteq B_n$, and so $\tilde{h}(\Delta) \subseteq \overline{B}_n$. Because $\tilde{h}(0) = av \in \partial B_n$ we conclude that $\tilde{h}$ is constant. Hence $h(\lambda) = \lambda a v$ for all λ, and so

$$g(\lambda) = \rho(h(\lambda)) = \rho(\lambda a v) = f(a\lambda),$$

as desired. This completes the proof of Theorem 2.5.

3. UNIQUENESS OF RIEMANN MAPPINGS, AND RIEMANN MAPPINGS ONTO CIRCLED DOMAINS

THEOREM 3.1. *(a) Suppose that* $\rho : B_n \to \mathbf{C}^n$ *and* $\tilde{\rho} : B_n \to \mathbf{C}^n$ *satisfy* (1.1)–(1.4) *and* $\rho(B_n) = \tilde{\rho}(B_n)$. *Then* $\alpha = \rho^{-1} \circ \tilde{\rho}$ *satisfies* (1.1)–(1.4).

(b) Conversely, if $\rho : B_n \to \mathbf{C}^n$ *and* $\alpha : B_n \to \mathbf{C}^n$ *satisfy* (1.1)–(1.4), *and if* $\alpha(B_n) = B_n$, *then* $\tilde{\rho} = \rho \circ \alpha$ *satisfies* (1.1)–(1.4).

This reduces the uniqueness question to the problem of characterizing the maps of B_n to itself for which (1.1)–(1.4) hold. The next result provides this characterization, and deals more generally with mappings whose image is completely circled. Before stating it we need a few definitions.

Let I be a completely circled domain. Define F_I on $\mathbf{C}^n$ by

$$F_I(z) = \inf\{r^2 : r > 0,\ z \in r\overline{I}\}, \tag{3.2}$$

where $r\overline{I} = \{rv : v \in \overline{I}\}$. Thus $F_I < 1$ on I, $F_I = 1$ on ∂I, and $F_I(\lambda z) = |\lambda|^2 F_I(z)$ for all $\lambda \in \mathbf{C}$, $z \in \mathbf{C}^n$. Let u_I be the function on $\overline{I}$ given by $\frac{1}{2}\log F_I$, and set $\omega_I = \frac{1}{2i}\partial\overline{\partial}F_I$. When $I = B_n$, these objects are simply F_0, u_0, and ω_0. Although ω_I is a priori only a current, it will usually be better than that in the circumstances that will occur here.

A mapping ϕ from a completely circled set in $\mathbf{C}^n$ to a complex vector space (such as $\mathbf{C}$, $\mathbf{C}^n$) will be called complex homogeneous of degree 1 if $\phi(\lambda z) = \lambda\phi(z)$ for all $\lambda \in \overline{\Delta}$, z in the domain of ϕ.

THEOREM 3.3. *Let I, etc., be as above. Suppose that* $\rho : B_n \to \mathbf{C}^n$ *satisfies* (1.1), (1.2), *and* $\rho(B_n) = I$. *Then ρ satisfies* (1.3) *and* (1.4) *if and only if ρ is complex homogeneous of degree* 1 *and* $\delta(\rho^{-1})(\omega_0) = \omega_I$ *on* $I \setminus \{0\}$.

Thus the ρ's for which (1.1)–(1.4) hold and $\rho(B_n) = B_n$ form a group. One way to produce elements of this group is to flow along vector fields on $\overline{B}_n$ that are tangent to ∂B_n on ∂B_n, complex homogeneous of degree 1, and which are Hamiltonian with respect to the symplectic form ω_0. More generally we have the following.

THEOREM 3.4. *Suppose that I is a completely circled, smooth, strongly pseudoconvex domain. Then there exists a map* $\rho : B_n \to \mathbf{C}^n$ *such that* (1.1)–(1.4) *hold,* $\rho(B_n) = I$, *and ρ is smooth away from the origin.*

The rest of this section will be spent proving these theorems.

Let ρ, $\tilde{\rho}$, α be as in (a) of Theorem 3.1. Clearly α satisfies (1.1) and (1.2). To check (1.3) and (1.4) we compute a little. Set $F = F_0 \circ \rho^{-1}$, $\widetilde{F} = F_0 \circ \tilde{\rho}^{-1}$, as in Section 2. Then $F = \widetilde{F}$, because of Theorem 2.3(e), and so $F_0 = F_0 \circ \alpha^{-1}$. We also have $\delta\alpha(\partial F_0) = \partial F_0$, because ρ, $\tilde{\rho}$ satisfy Theorem 2.2(b), and so (1.3) and (1.4) hold for α, by Theorem 2.2.

Suppose now that ρ, α, $\tilde{\rho}$ are as in (b) of Theorem 3.1. The argument is almost the same as before. Theorem 2.3(e) gives $F_0 \circ \alpha^{-1} = F_0$, so that $F_0 \circ \tilde{\rho}^{-1} = F_0 \circ \rho^{-1}$, which we denote by F as usual. Applying Theorem 2.2 to ρ, α we get that $\delta\tilde{\rho}(\partial F) = \partial F_0$, and since $\tilde{\rho}$ clearly satisfies (1.1) and (1.2) if ρ, α do, we obtain from Theorem 2.2 that (1.3) and (1.4) also hold for $\tilde{\rho}$.

Before proving Theorem 3.3 we need a preliminary fact.

LEMMA 3.5. *Suppose that I is completely circled and pseudoconvex. Then u_I is plurisubharmonic and it is the Green's function for I.*

This is well known, but we give a proof for completeness.

We first show that u_I is plurisubharmonic. It suffices to show that if f is any holomorphic map of Δ into I, then

$$u_I(f(0)) \leq \frac{1}{2\pi}\int_0^{2\pi} u_I(f(re^{i\theta}))d\theta$$

for all $r \in (0,1)$. Fix f, r. It is easy to reduce to the case where $f(0) \neq 0$, and $f(re^{i\theta})$ is nonzero for each θ.

Let h be the real-valued harmonic function on $\Delta_r = \{\lambda \in \Delta : |\lambda| < r\}$ such that $h = u_I \circ f$ on $\partial\Delta_r$. Let $\tilde{h}$ denote the harmonic conjugate of h on Δ_r that vanishes at 0, and set $H = \exp(-(h + i\tilde{h}))$. Define $g : \overline{\Delta}_r \to \mathbf{C}^n$ by $g = Hf$, so that g is holomorphic on Δ_r and g maps $\partial\Delta_r$ into $\overline{I}$. This last is easily checked by chasing definitions.

We must also have that $g(\overline{\Delta}_r) \subseteq \overline{I}$. Indeed, let R be the smallest positive number such that $g(\overline{\Delta}_r) \subseteq R\overline{I}$, and suppose that $R > 1$. This would contradict the pseudoconvexity of RI, since $g(\partial\Delta_r) \subseteq \overline{I} \subseteq RI$. Thus $g(\overline{\Delta}_r) \subseteq \overline{I}$, and hence $u_I \circ g \leq 0$ on $\overline{\Delta}_r$. This implies that $u_I \circ f \leq h$ on $\overline{\Delta}_r$, so that

$$u_I(f(0)) \leq h(0) = \frac{1}{2\pi}\int_0^{2\pi} h(re^{i\theta})d\theta = \frac{1}{2\pi}\int_0^{2\pi} u_I(f(re^{i\theta}))d\theta,$$

as desired. Thus u_I is plurisubharmonic.

To show that u_I is the Green's function on I we must show that u_I is given by (2.4), with D replaced by I. Clearly u_I is a competitor for the supremum in (2.4), and so we are reduced to showing that $u_I \geq v$ for all v as in (2.4).

Fix $z \in \partial I$. Then $v(\lambda z)$ defines a nonpositive subharmonic function in λ that is $\leq \log|\lambda| + C$ for some C that depends only on v. This forces $v(\lambda z) \leq \log|\lambda| = u_I(\lambda z)$. Since z is arbitrary we conclude that $v \leq u$. This proves the lemma.

Let us now prove Theorem 3.3. Suppose first that ρ satisfies (1.1)–(1.4). Then Theorem 2.3(e) and Lemma 3.5 imply that $F_I = F_0 \circ \rho^{-1}$, and so Lemma 2.6 gives that ω_I is continuous and $\delta(\rho^{-1})(\omega_0) = \omega_I$. To show that ρ is complex homogeneous of degree 1 we will use ω_0, ω_I to convert $\delta\rho(\partial F_I) = \partial F_0$ into a statement about vector fields.

Before entering into the computations we should make a remark about vector fields on $\mathbf{C}^n$ that will be relevant elsewhere in this paper. When we refer to a vector field on $\mathbf{C}^n$, or on a subdomain of $\mathbf{C}^n$, we mean a vector field in the usual sense of smooth manifolds, ignoring the complex structure. We shall, however, usually represent such a vector field V by its complex components V_k, so that

$$V(f) = \Sigma V_k \partial_k f + \Sigma \overline{V}_k \overline{\partial}_k f \tag{3.6}$$

for all functions f.

Let V, W be two smooth vector fields on $\mathbf{C}^n$ with (complex) components V_k and W_k. Then on $I \setminus \{0\}$ we have

$$\begin{aligned} i(V)(\omega_I)(W) = \omega_I(V, W) &= \frac{1}{2i}\sum_{j,k}\left(\frac{\partial^2}{\partial z_j \partial \overline{z}_k} F_I\right)(V_j \overline{W}_k - W_j \overline{V}_k) \\ &= \operatorname{Im}\left\{\sum_{j,k}\left(\frac{\partial^2}{\partial z_j \partial \overline{z}_k} F_I\right) V_j \overline{W}_k\right\}. \end{aligned} \tag{3.7}$$

If we take V to be the vector field with $V_k = iz_k$, then we have

$$\begin{aligned} i(V)(\omega_I)(W) &= \operatorname{Re}\left\{\sum_{j,k}\left(\frac{\partial^2}{\partial z_j \partial \overline{z}_k} F_I\right) z_j \overline{W}_k\right\} \\ &= \operatorname{Re}\left\{\sum_{k}\left(\frac{\partial}{\partial \overline{z}_k} F_I\right) \overline{W}_k\right\} = \frac{1}{2} dF_I(W). \end{aligned} \tag{3.8}$$

In the second equality we used the homogeneity identity

$$\begin{aligned} \sum_j \frac{\partial^2}{\partial z_j \partial \overline{z}_k} F_I(z) z_j &= \frac{\partial}{\partial \lambda}\bigg|_{\lambda=1}\left(\frac{\partial}{\partial \overline{z}_k} F_I(\lambda z)\right) \\ &= \frac{\partial}{\partial \lambda}\bigg|_{\lambda=1}\left[\lambda\left(\frac{\partial}{\partial \overline{z}_k} F_I(z)\right)\right] = \frac{\partial}{\partial \overline{z}_k} F_I(z). \end{aligned} \tag{3.9}$$

The second equality can be derived from $F_I(\lambda z) = |\lambda|^2 F_I(z)$. Strictly speaking, we should view these formulae as holding in the sense of distributions, but it is easy to reduce to the C^2 case by approximation.

We can now conclude that $d\rho(V) = V$. This uses (3.8), the corresponding calculation for ω_0, and also $\delta(\rho^{-1})(\omega_0) = \omega_I$, $\delta(\rho^{-1})(dF_0) = dF_I$, and the non-degeneracy of ω_0, ω_I. Hence $\rho(e^{it}z) = e^{it}\rho(z)$ for all $t \in \mathbf{R}$, $z \in B_n$.

Next we apply (3.7) to V defined by $V_k = z_k$. This time we get that

$$i(V)(\omega_I)(W) = \operatorname{Im}(\overline{\partial} F_I(W)).$$

Arguing as above we obtain that $d\rho(V) = V$, whence $\rho(e^t z) = e^t \rho(z)$ for $z \in B_n$, $t \leq 0$. This proves the "only if" part of Theorem 3.3.

The "if" part is established using the same arguments, but in reverse. Suppose that ρ is complex homogeneous of degree 1, that (1.1) and (1.2) hold, and that $\delta(\rho^{-1})(\omega_0) = \omega_I$ on $I \setminus \{0\}$, so that in particular ω_I is continuous away from the origin. Then $d\rho(V) = V$ for the two choices of V as above, and the same calculations as before now yield $\delta(\rho^{-1})(\partial F_0) = \partial F_I$. From Theorem 2.2 ((c) implies (a)) we conclude that ρ satisfies (1.2) and (1.4).

The argument we use to prove Theorem 3.4 is essentially the same as used by Moser in [M] to produce mappings with prescribed effect on a symplectic form. (See also [We].) In our situation homogeneity will substitute for compactness of the domain manifold.

In what follows we shall always view F_I, F_0 as being defined and homogeneous on all of $\mathbf{C}^n$, not just on I, B_n. The smoothness and strong pseudoconvexity of I imply that F_I is smooth and strictly plurisubharmonic away from the origin.

We are going to obtain ρ by flowing along a vector field. To do this we need to join B_n to I by a curve of completely circled, smooth, strongly pseudoconvex domains, but it doesn't matter how we do this, and so we'll do it in the way that is simplest computationally.

Set $F_t = (1-t)F_0 + tF_I$, $0 \leq t \leq 1$, so that each F_t is smooth and strictly plurisubharmonic away from the origin, and each F_t satisfies $F_t(\lambda z) = |\lambda|^2 F_t(z)$ for all $\lambda \in \mathbf{C}$, $z \in \mathbf{C}^n$. Let V_t be the vector field on $\mathbf{C}^n$ whose complex components (in the sense of (3.6)) are given by

$$V_{t,k} = -\frac{1}{2} \sum_l F_t^{k\bar{l}} \left(\frac{\partial}{\partial \bar{z}_l} \dot{F}_t \right),$$

where the dot denotes a derivative in t, and $F_t^{k\bar{l}}$ is the matrix inverse to $\frac{\partial^2}{\partial z_k \partial \bar{z}_l} F_t$, i.e.,

$$\sum_l F_t^{k\bar{l}} \left(\frac{\partial^2}{\partial z_j \partial \bar{z}_l} F_t \right) = \delta_j^k .$$

We want to obtain ρ by flowing along V_t. To show that this works we have to check a few things. The first thing we want to do is compute $V_t(F_t)$ to show that the flow preserves F_t:

$$\begin{aligned}
(3.10) \qquad V_t(F_t) &= -\operatorname{Re}\left\{ \sum_{k,l} F_t^{k\bar{l}} \left(\frac{\partial}{\partial \bar{z}_l} \dot{F}_t \right) \left(\frac{\partial}{\partial z_k} F_t \right) \right\} \\
&= -\operatorname{Re}\left\{ \sum_{k,l,m} F_t^{k\bar{l}} \left(\frac{\partial}{\partial \bar{z}_l} \dot{F}_t \right) \left(\frac{\partial^2}{\partial z_k \partial \bar{z}_m} F_t \right) \bar{z}_m \right\} \\
&= -\operatorname{Re}\left\{ \sum_l \frac{\partial}{\partial \bar{z}_l} \dot{F}_t \bar{z}_l \right\} = -\dot{F}_t .
\end{aligned}$$

We used the complex conjugate of (3.9) (with different indices) in the second equality, while in the last we used the homogeneity identity

(3.11)

$$\sum_l \left(\frac{\partial}{\partial \overline{z}_l}\dot{F}_t\right)(z)\overline{z}_l = \frac{\partial}{\partial \overline{\lambda}}\Big|_{\lambda=1}\left(\dot{F}_t(\lambda z)\right) = \frac{\partial}{\partial \overline{\lambda}}\Big|_{\lambda=1}\left(|\lambda|^2\dot{F}_t(z)\right) = \dot{F}_t(z).$$

From (3.10) it follows that if $\rho_t(z)$ is a flow on $\mathbf{C}^n$ that satisfies

$$\dot{\rho}_t(z) = V_t(\rho_t(z)), \tag{3.12}$$

then $F_t(\rho_t(z))$ is constant in t. This implies that we can solve (3.12) on all of $\mathbf{C}^n \times [0,1]$ with initial data $\rho_t = \mathrm{id}$, that is, there is no blow-up.

Set $\omega_t = \frac{1}{2i}\partial\overline{\partial}F_t$. It is not too hard to calculate that

$$L_{V_t}(\omega_t) = -\dot{\omega}_t, \tag{3.13}$$

where L_V denotes the Lie derivative. This is a special case of (3.14) in Proposition 3.11 in [S]. It implies that $\delta\rho_t(\omega_t) = \omega_0$.

Take $\rho = \rho_1 \big|_{B_n}$. Then $\delta\rho(\omega_I) = \omega_0$ and $F_I \circ \rho = F_0$, and in particular ρ maps B_n onto I. It is not hard to verify that ρ is complex homogeneous of degree 1, because of the corresponding property for V_t. It is also not hard to see that ρ satisfies (1.1) and (1.2), and that ρ is smooth away from the origin. Theorem 3.4 now follows from Theorem 3.2.

Note that the ρ produced by the argument above is real-analytic away from the origin when ∂I is real analytic. We could also weaken the regularity conditions on ∂I, but we have to be careful, because this approach suffers a loss of derivatives; to get ρ to be C^1 away from the origin, we need F_I to be C^3 away from 0.

4. RIEMANN MAPPINGS AND THE KOBAYASHI INDICATRIX

Let D be a bounded domain in $\mathbf{C}^n$ that contains the origin. Let I denote its Kobayashi indicatrix, which is the domain defined by

$$I = \{v \in \mathbf{C}^n : v = f'(0) \text{ for some holomorphic map } f : \Delta \to D \text{ such that}$$
$$f(0) = 0 \text{ and } \overline{f(\Delta)} \subseteq D\}. \tag{4.1}$$

Notice that I is always completely circled.

THEOREM 4.2. *Suppose that $\rho : B_n \to \mathbf{C}^n$ satisfies (1.1)–(1.4) and $\rho(B_n) = D$, and let I be as in (4.1). Define $\sigma : B_n \to \mathbf{C}^n$ by*

$$\sigma(z) = \frac{\partial}{\partial \lambda}\bigg|_{\lambda=0} (\rho(\lambda z)) \qquad \text{for } z \in B_n. \tag{4.3}$$

Then σ satisfies (1.1)–(1.4), $\sigma(B_n) = I$, and σ is complex homogeneous of degree 1.

It is clear that (4.3) makes sense, because (1.3) holds for ρ, and that σ is complex homogeneous of degree 1. This implies in particular that σ satisfies (1.3).

To verify the other conclusions about σ it is helpful to write down a couple of other formulae for σ. For example, we can rewrite (4.3) as

$$\sigma(z) = \lim_{t \to 0} \frac{\rho(tz)}{t}, \tag{4.4}$$

from which it follows easily that σ is a bilipschitz map of B_n onto its image, since ρ is bilipschitz on $\frac{1}{2}B_n$ (and $\rho(0) = 0$).

Next we rewrite (4.3) in terms of a Cauchy integral formula. Because ρ satisfies (1.3) we have

$$\rho(\lambda z) = \frac{1}{2\pi i} \int_{|\alpha|=1} \frac{\rho(\alpha z)}{\alpha - \lambda} d\alpha \tag{4.5}$$

for all $z \in B_n$, $\lambda \in \Delta$, and hence

$$\sigma(z) = \frac{1}{2\pi i} \int_{|\alpha|=1} \frac{\rho(\alpha z)}{\alpha^2} d\alpha. \tag{4.6}$$

This implies that σ is C^1 away from the origin, which, combined with the preceeding observations, gives us (1.1) and (1.2) for σ.

Using $\rho(0) = 0$ we also have

$$\rho(\lambda z) = \frac{1}{2\pi i} \int_{|\alpha|=\frac{1}{2}} \rho(\alpha z) \left\{ \frac{1}{\alpha - \lambda} - \frac{1}{\alpha} \right\} d\alpha = \frac{1}{2\pi i} \int_{|\alpha|=\frac{1}{2}} \rho(\alpha z) \frac{\lambda}{(\alpha - \lambda)\alpha} d\alpha. \tag{4.7}$$

for $\lambda \in \Delta$, $|\lambda| < \frac{1}{2}$. Hence

$$\lambda^{-1}\rho(\lambda z) - \sigma(z) = \frac{1}{2\pi i} \int_{|\alpha|=\frac{1}{2}} \rho(\alpha z) \frac{\lambda}{(\alpha - \lambda)\alpha^2} d\alpha. \tag{4.8}$$

This implies that $\lambda^{-1}\rho(\lambda z)$ converges to $\sigma(z)$ as $\lambda \to 0$ in the C^1 topology on $B_n \setminus \{0\}$. Using this it is not hard to show that σ satisfies (1.4), since ρ does.

It remains to verify that $\sigma(B_n) = I$. This follows from Theorem 2.5 and the fact that σ is bilipschitz. Indeed, the latter and $\sigma(0) = 0$ force the image of σ to contain a neighborhood of the origin. Hence for each $w \in \mathbf{C}^n$ there is a $v \in \partial B_n$ so that $\frac{\partial}{\partial\lambda}\big|_{\lambda=0} \rho(\lambda v) = aw$ for some $a \in \mathbf{C}$, $a \neq 0$. If $w \in I$, then $|a| < 1$, by Theorem 2.5, and so $\sigma(B_n) \supseteq I$. This does the job, since $\sigma(B_n) \subseteq I$ follows from the definitions.

5. EXISTENCE OF RIEMANN MAPPINGS WHOSE IMAGE IS A GIVEN SMOOTH, STRONGLY CONVEX DOMAIN

Throughout this section we let D be a (fixed) domain in $\mathbf{C}^n$ that is smooth, strongly convex, and which contains the origin. We are going to use the work of Lempert [L1, **3**] to show that there is a $\rho : B_n \to \mathbf{C}^n$ that satisfies (1.1)–(1.4) and which extends to a smooth diffeomorphism of $\overline{B}_n \setminus \{0\}$ onto $\overline{D} \setminus \{0\}$.

Let I be the Kobayashi indicatrix of D. As in Section 2 of [L3] I has smooth boundary (because of [L1]) and is strongly convex in addition to being completely circled.

From [L1] we know that for each $v \in \partial I$ there is a unique holomorphic map $f_v : \Delta \to D$ such that $f_v(0) = 0$, $f_v'(0) = v$. We shall refer to these maps as "extremal mappings." They extend to smooth embeddings of $\overline{\Delta}$ into $\overline{D}$ that send $\partial\Delta$ into ∂D, and they depend smoothly on v.

Define $\Psi : \overline{I} \to \overline{D}$ by

$$\Psi(rv) = f_v(r) \text{ when } v \in \partial I,\ r \in [0,1]. \tag{5.1}$$

In [L1] it is shown that Ψ is a homeomorphism of $\overline{I}$ onto $\overline{D}$ and a C^∞ diffeomorphism away from the origin.

THEOREM 5.2. *Suppose that $\sigma : B_n \to \mathbf{C}^n$ satisfies* (1.1)–(1.4) *and $\sigma(B_n) = I$. Then $\rho = \Psi \circ \sigma$ satisfies* (1.1)–(1.4).

Recall that Theorem 3.4 guarantees the existence of such a σ, even one that is smooth away from the origin.

We only need to show that ρ satisfies (1.4) and that Ψ is bilipschitz on $\overline{I}$; the rest follows from what we already know about Ψ.

To check that (1.4) holds for ρ we need some notation. Define subbundles $S^1(I)$, $S^2(I)$ of the tangent bundle of $I \setminus \{0\}$ as follows. Suppose that $z \in I \setminus \{0\}$, and choose $r \in (0,1)$ so that $z \in \partial(rI)$. Let $S_z^1(I)$ be the complex line through z, and let $S_z^2(I)$ be the maximal complex subspace of the tangent space at z of the boundary of rI. Then $S_z^1(I) + S_z^2(I) = \mathbf{C}^n$, and $S^1(B_n)$, $S^2(B_n)$ are the same as the S^1, S^2 defined in Section 1.

It follows from the arguments in Section 5 of [L3] that $d\Psi_z$ maps $S_z^2(I)$ to a complex subspace for each $z \in I \setminus \{0\}$. Therefore, in order to show that ρ satisfies (1.4), it is enough to show that $d\sigma$ sends $S^2(B_n)$ to $S^2(I)$.

We know that $F_I \circ \sigma = F_0$, because of Theorem 2.3(e) and Lemma 3.5, and hence $\delta\sigma(\partial F_I) = \partial F_0$, by Theorem 2.2. It is easy to check that $S_z^2(I)$ is the kernel of $\partial F_I \big|_z$, and similarly for $S^2(B_n)$ and ∂F_0. Hence $d\sigma$ sends $S^2(B_n)$ to $S^2(I)$, as desired.

Next we check that Ψ is Lipschitz. We know that it is smooth away from the origin, and so we only need to look at its behavior near 0. As in (4.7) we get from the Cauchy integral formula and $\Psi(0) = 0$ that

$$\Psi(\lambda z) = \frac{1}{2\pi i} \int\limits_{|\alpha|=1} \Psi(\alpha z) \frac{\lambda}{(\alpha - \lambda)\alpha} d\alpha \tag{5.3}$$

for $z \in \overline{I}$, $\lambda \in \Delta$. Using this (with $z \in \partial I$, $\lambda \in [0,1)$) and the smoothness of Ψ away from the origin it is easy to show that Ψ is Lipschitz on a neighborhood of the origin, and hence on all of $\overline{I}$.

It remains to verify that Ψ is bilipschitz on $\overline{I}$. Because Ψ is a homeomorphism on $\overline{I}$ and a diffeomorphism away from the origin we need only check this on a neighborhood of the origin.

From the definition of Ψ we know that $\frac{\partial}{\partial \lambda} \big|_{\lambda=0} \Psi(\lambda z) = z$. As in (4.8) we therefore have

$$\lambda^{-1}\Psi(\lambda z) - z = \frac{1}{2\pi i} \int\limits_{|\alpha|=1} \Psi(\alpha z) \frac{\lambda}{(\lambda - \alpha)\alpha^2} d\alpha. \tag{5.4}$$

Because Ψ is Lipschitz, the Lipschitz norm of the right hand side on $\overline{I}$ tends to zero as $\lambda \to 0$. This implies that $\lambda^{-1}\Psi(\lambda z)$ is bilipschitz on $\overline{I}$ if λ is small enough, which yields the bilipschitzness of Ψ on a neighborhood of the origin. This proves Theorem 5.2.

6. RIEMANN MAPPINGS AND HCMA, PART 1

If $n = 1$ and ρ is a Riemann mapping, then $f \mapsto f \circ \rho$ takes harmonic functions to harmonic functions. When $n > 1$, $f \mapsto f \circ \rho$ takes solutions of HCMA to solutions of HCMA when ρ is holomorphic, but not for a general Riemann mapping. However, it turns out that Riemann mappings — under some extra real-analyticity assumptions — induce an action on functions that is nonlinear in general but that does send solutions of HCMA to other solutions. This action is a special case of a more general phenomenon discovered by Lempert [L4], except for some minor modifications needed for analytical reasons (having to do with lack of regularity at the origin). In this section we describe Lempert's method and the required modifications, and we implement it in our situation in the next section.

In rough terms the basic idea behind Lempert's method is as follows. Given a real-valued function A on a domain $\mathbf{C}^n$, we can look at the graph of ∂A in the complex cotangent bundle of $\mathbf{C}^n$. It turns out that the graphs that arise from solutions of HCMA can be characterized in terms of the natural holomorphic symplectic structure on the complex cotangent bundle. In particular this characterization is invariant under symplectic biholomorphisms.

Let us make this precise. Let $\mathcal{C} = \mathbf{C}^n \times \mathbf{C}^n$. We identify this with the complex cotangent bundle of $\mathbf{C}^n$; given $(z, \zeta) \in \mathcal{C}$, we associate to ζ the $(1,0)$-form $\Sigma \zeta_j dz_j$ at z, and conversely, given $z \in \mathbf{C}^n$ and a $(1,0)$-form at z, we get a point (z, ζ) in $\mathcal{C}$. Define a holomorphic symplectic form γ on $\mathcal{C}$ by

$$\gamma = \frac{1}{2i} \sum_j dz_j \wedge d\zeta_j, \tag{6.1}$$

where $\{z_j, \zeta_j\}$ are the obvious co-ordinates on $\mathcal{C}$. Define also

$$\Gamma = \sum \zeta_j dz_j, \tag{6.2}$$

so that

$$\gamma = -\frac{1}{2i} d\Gamma. \tag{6.3}$$

Let μ, ν denote the real and imaginary parts of γ, each of which is a symplectic form on $\mathcal{C}$ in its own right.

We need to record some formulae concerning graphs in $\mathcal{C}$. Let β be a $\mathbf{C}^n$-valued function on some region R in $\mathbf{C}^n$, and define $\tau : R \to \mathcal{C}$ by

$$\tau(z) = (z, \beta(z)). \tag{6.4}$$

Clearly then

$$\delta\tau(\Gamma) = \sum \beta_j dz_j, \tag{6.5}$$

and so the graph N of β is ν-Lagrangian if and only if

$$\delta\tau(\nu) = d\left(\operatorname{Re}\frac{1}{2}\sum \beta_j dz_j\right) \tag{6.6}$$

vanishes. [Recall that N is ν-Lagrangian if the real dimension of N is half that of the total space, and if $\delta i_N(\nu) = 0$, where $i_N : N \to \mathcal{C}$ is the usual inclusion map.]

Set $\alpha = 2\operatorname{Re}\sum \beta_j dz_j$, so that N is ν-Lagrangian iff $d\alpha = 0$. Assume for the rest of this section that R has trivial first deRham cohomology, so that there is a real-valued function A such that $\alpha = dA$. In this case we have that $\beta_j = \partial_j A$, so that $\sum \beta_j dz_j = \partial A$ is ∂-closed.

If N is ν-Lagrangian, and A is as above, then

$$\delta\tau(\mu) = \delta\tau(\gamma) = -\frac{1}{2i}d(\delta\tau(\Gamma)) = -\frac{1}{2i}d(\partial A) = \frac{1}{2i}\partial\bar{\partial}A. \tag{6.7}$$

This implies the following result. Set $\mu_N = \delta i_N(\mu)$.

THEOREM 6.8. ([**L4**]) *Let N be the graph of β as above. Then the following are equivalent:*

(a) *there is a real-valued function A such that $\beta_j = \partial_j A$ and A satisfies HCMA;*
(b) *N is ν-Lagrangian and $\mu_N^n = 0$.*

Condition (b) is invariant under holomorphic symplectomorphisms. If U is an open set in $\mathcal{C}$ and $g : U \to \mathcal{C}$ is a biholomorphic γ-symplectomorphism onto its image, and if $N \subseteq U$, then $g(N)$ satisfies (b) when N does. Of course it need not be true that $g(N)$ is a graph if N is, although the set of graphs that get taken to graphs is open in the C^1 topology (as long as you stay away from ∂U). Also, if N_1, N_2 are two graphs over the same region, that may not be true of $g(N_1)$, $g(N_2)$, even if they are both graphs, i.e., the domains can change.

An important class of examples of g's comes from biholomorphisms on $\mathbf{C}^n$. If D_1, D_2 are two domains in $\mathbf{C}^n$, and $f : D_1 \to D_2$ is a biholomorphism, then

$$g((z,\zeta)) = (f(z), (df_z)^{-t}(\zeta)) \tag{6.9}$$

is a biholomorphism of $D_1 \times \mathbf{C}^n \subseteq \mathcal{C}$ onto $D_2 \times \mathbf{C}^n \subseteq \mathcal{C}$. Here $(df_z)^{-t}$ denotes the inverse transpose of df_z, where the transpose is defined via the representation of df_z as a complex matrix, using the standard basis on $\mathbf{C}^n$. It is easy to check that for this choice of g we have $\delta g(\Gamma) = \Gamma$, $\delta g(\gamma) = \gamma$. Also, if N is the graph of ∂A, then $g(N)$ is the graph of $\partial(A \circ g^{-1})$.

Another way to produce holomorphic symplectomorphisms g is to flow along holomorphic symplectic vector fields. A third approach, which is closer to the concerns of this paper, goes as follows. Let D_1, D_2 be two domains in $\mathbf{C}^n$, and let N_1, N_2 be two graphs in $\mathcal{C}$ sitting over D_1, D_2. Let $f : D_1 \to D_2$ be some mapping, which we lift to a map $\hat{f} : N_1 \to N_2$. If f, N_1, N_2 are real-analytic, and if N_1 is totally real, then $\hat{f}$ can be extended to a holomorphic map g on a neighborhood of N_1. If $\delta i_{N_1}(\delta g(\gamma)) = \delta \hat{f}(\gamma)$ equals $\delta i_{N_1}(\gamma)$, then $\delta g(\gamma) = \gamma$, because N_1 is totally real.

Although this third approach is pretty much the sort of thing we want and will use, there are some problems with it. The first is that the domain of g may be a small neighborhood around N_1, and so if you want g to act on graphs of gradients of functions that satisfy HCMA, it would be good if N_1 were already such a graph. However, this would be incompatible with N_1 being totally real; if N_1 is ν-Lagrangian, then it is easy to see that it is totally real if and only if μ_N is nondegenerate.

This is not as bad as it may sound. Suppose that N_1 is the graph of ∂A_1, where A_1 satisfies HCMA, but $(\partial\overline{\partial} A_1)^{n-1}$ is never zero. This means that N_1 misses being totally real by only one direction, and so we might be able to get a holomorphic extension g of f if we impose further conditions on f to take care of the bad part. For example, we know from [BK] that there is a foliation of D_1 by Riemann surfaces associated to A_1, and it is not hard to show that the map $\tau_1 : D_1 \to N_1$ defined as in (6.4) is holomorphic on these surfaces. If we demand that f also be holomorphic on them, then we can try to use that to deal with the bad directions.

There is another problem which is more serious. We shall eventually be working with solutions of HCMA provided by [L1, **2**], and they have a logarithmic singularity at the origin. To deal with this it will be preferable to work with $F = \exp(2u)$ instead of u itself, where u is a function that satisfies HCMA. This is also convenient for dealing with the other problem; $\partial\overline{\partial} F$ will be nondegenerate in the cases we consider, and so the graph of ∂F will be totally real. However, we now have the problem of encoding the fact that $\log F$ satisfies HCMA in terms of the graph of ∂F. It is not clear how to do this using only γ, but we can do it if we use γ and Γ.

THEOREM 6.10. *Let N be the graph of β, as before. The following are equivalent:*

(a) *β can be represented as $\beta_j = \partial_j A$, where A is a positive function such that $\log A$ satisfies HCMA;*
(b) *N is ν-Lagrangian, and there is a positive function B on N such that $dB = 2\,\mathrm{Re}\,\delta i_N(\Gamma)$ and*

$$\left[\mu_N - \frac{1}{2i} B^{-1} \delta i_N(\Gamma) \wedge \delta i_N(\overline{\Gamma})\right]^n = 0.$$

Notice that $\mathrm{Re}\,\delta i_N(\Gamma)$ is closed when N is ν-Lagrangian.

The proof of this is pretty much the same as for Theorem 6.8. In proving that (a) implies (b) we take $B = A \circ \tau^{-1}$, where τ is as in (6.4), and for the converse we set $A = B \circ \tau$. Also, from (6.5) and (6.7) we have that

(6.11)
$$\delta\tau\left(\mu_N - \frac{1}{2i}B^{-1}\delta i_N(\Gamma)\wedge\delta i_N(\overline{\Gamma})\right) = \frac{1}{2i}\partial\overline{\partial}A - \frac{1}{2i}A^{-1}\partial A\wedge\overline{\partial}A = \frac{1}{2i}A(\partial\overline{\partial}\log A),$$

which shows that the last condition in (b) corresponds to $\log A$ satisfying HCMA.

Analogous to the previous situation we have that if U is an open subset of $\mathcal{C}$ and $g : U \to \mathcal{C}$ is a biholomorphism of U onto its image, and if $\delta g(\gamma) = \gamma$, $\delta g(\Gamma) = \Gamma$, and $N \subseteq U$, then $g(N)$ satisfies (b) when N does.

We can also produce maps g with these properties in much the same way as before. We already observed that a biholomorphism f of one domain onto another gives rise to such a g by (6.9). We can again flow along a holomorphic vector field V, but now we have to require that $L_V\Gamma = 0$, where L_V denotes the Lie derivative, and not just that V is γ-symplectic. There is a version of the third approach, which works as follows.

Let D_1, D_2 be domains in $\mathbf{C}^n$, let $f : D_1 \to D_2$ be real analytic, and let N_1, N_2 be two real-analytic graphs in $\mathcal{C}$ over D_1, D_2. It is now reasonable to require that N_1 be totally real, and so the lift $\hat{f} : N_1 \to N_2$ extends to a holomorphic map g on a neighborhood of N_1. If now $\delta\hat{f}(\Gamma) = \delta i_{N_1}(\Gamma)$, then $\delta\hat{f}(\gamma) = \delta i_{N_1}(\gamma)$ also, and we have $\delta g(\Gamma) = \Gamma$, $\delta g(\gamma) = \gamma$ because N_1 is totally real. The difference between this and what we did before is that we now require that $\delta\hat{f}(\Gamma) = \delta i_{N_1}(\Gamma)$, but we don't have to worry about the "bad direction." However, for N_1 as in Proposition 6.10, the extra condition on $\delta\hat{f}(\Gamma)$ really amounts to taking care of the bad direction.

We shall implement this third approach for producing g's in our situation in the next section.

Before leaving this section there is an observation that should be recorded. Let U be an open subset of $\mathcal{C}$, and $\phi : U \to \mathcal{C}$ be a ν-symplectomorphism onto its image. Let D_1, D_2 be two domains in $\mathbf{C}^n$, let A_1, A_2 be two real-valued functions on D_1, D_2, and let N_1, N_2 denote the graphs of ∂A_1, ∂A_2. Assume that $\overline{N_1} \subseteq U$, and that $\phi(N_1) = N_2$. Let us also make the normalizing assumptions that $0 \in D_1$, D_2, $A_1(0) = A_2(0) = 0$, and that $\phi((0, \partial A_1(0))) = (0, \partial A_2(0))$. If A is a function on D_1 such that $A - A_1$ has small C^2 norm, say, and $A(0) = 0$, then the graph N of ∂A also satisfies $\overline{N} \subseteq U$, and $g(N)$ is the graph of ∂ applied to a function $\Phi(A)$ which vanishes at 0. The domain of $\Phi(A)$ need not be D_2, but it will be close to D_2 when $A - A_1$ is small. Then the differential of Φ is given by composition, i.e.,

$$\frac{d}{dt}\Big|_{t=0}(\Phi(A_1 + ta_1)) = a_1 \circ \psi^{-1}, \tag{6.12}$$

for all C^2 functions a_1 on $\overline{D_1}$ with $a_1(0) = 0$, where ψ is the map from D_1 to D_2 whose lifting $\hat{\psi} : N_1 \to N_2$ is equal to $\phi\,\big|_{N_1}$.

Before verifying (6.12) let us make some remarks. The left side of (6.12) is only intended to be defined pointwise, as a function on D_2, and does not take into account how the domains of $\Phi(A_1 + ta_1)$ are changing, although that could be computed. In other words, if you fix $z \in D_2$, then z will lie in the domain of $\Phi(A_1 + ta_1)$ for t small enough, and so the left side of (6.12) will be defined.

A nice thing about (6.12) is that it helps to make precise the idea that Φ is some kind of nonlinear generalization of a composition operator. This is related to the idea that if you want to find versions of the Riemann mapping theorem in other contexts, then it might be good to allow some more general notion of a mapping. For instance, if you think of a mapping as inducing an operation on functions by composition, then you could also look at operations that don't arise from composition. You'd still want to work with operations that contain some geometric information, like Φ, or Fourier integral operators.

Let us now check (6.12). We could use the fact that $(\mathcal{C}, \nu)$ is naturally isomorphic to $T^*\mathbf{C}^n$ with its canonical (real) symplectic structure, and compute everything in terms of $T^*\mathbf{C}^n$. Although it is more pleasant to work on $T^*\mathbf{C}^n$, it is not clear that keeping track of the isomorphism is easier than doing the analogous calculations on $(\mathcal{C}, \nu)$, and so we'll take the latter approach.

Let us first show that we can reduce to the case where $A_1, A_2 \equiv 0$. Define ρ_j mapping $D_j \times \mathbf{C}^n$ to itself, $j = 1, 2$, by

$$\rho_j((z, \zeta)) = (z, \zeta + \{\partial_k A_j(z)\}_k).$$

Then $\delta\rho_j(\Gamma) = \Gamma + \delta\Pi(\partial A_j)$, where $\Pi : \mathcal{C} \to \mathbf{C}^n$ is the projection onto the first set of co-ordinates. This implies that $\delta\rho_j(\nu) = d(\operatorname{Im}\delta\rho_j(-\frac{1}{2i}\Gamma)) = \nu$. Using the ρ_j's we can reduce to the case where A_1, A_2 are identically zero; otherwise we replace ϕ by $\rho_2^{-1} \circ \phi \circ \rho_1$.

Now, armed with the assumption that A_1 and A_2 are zero, we can pretty much compute directly. The argument that follows was suggested by the referee, and it is much simpler than the one it replaces. I am grateful to the referee for providing this improvement.

Set $a_2^t = \Phi(ta_1)$, $\beta_1 = \partial a_1$, and $\beta_2^t = \partial a_2^t$, so that $a_2^t \equiv 0$ at $t = 0$, by assumption. Let differentiation in t be denoted by a dot, and let the absence of the superscript t in some expression denote evaluation of this expression at $t = 0$. With this notation (6.12) becomes

$$\dot{a}_2 \circ \psi = a_1. \tag{6.13}$$

Because $a_1(0) = 0 = a_2^t(0)$, this is equivalent to showing that

$$V(\dot{a}_2 \circ \psi) = V(a_1) \tag{6.14}$$

for all tangent vectors V to D_1.

By definition of ϕ we have

$$\phi(z, t\beta_1(z)) = (\psi^t(z), \beta_2^t(\psi^t(z))) \tag{6.15}$$

for some 1-parameter family of mappings $\psi^t : D_1 \to \mathbf{C}^n$. (Note that $\psi^t \big|_{t=0}$ equals the mapping $\psi : D_1 \to D_2$ defined above, and so the notational conventions are consistent. Also, the notation is being slightly abused here; although the β's are $(1,0)$-forms, they are being identified with $\mathbf{C}^n$-valued functions in the obvious way.) Differentiation of (6.15) in t followed by evaluation at $t = 0$ yields

$$d\phi_{(z,0)}(0, \beta_1(z)) = (\dot{\psi}(z), \dot{\beta}_2(\psi(z))). \tag{6.16}$$

This uses the fact that $\beta_2^t \equiv 0$ when $t = 0$.

We want to use the assumption that ϕ is a symplectomorphism to deduce (6.14) from (6.16). Let $\beta_{1,j}(z)$ and $\beta_{2,j}^t(z)$, $1 \leq j \leq n$, denote the components of $\beta_1(z)$ and $\beta_2^t(z)$. Fix $z \in D_1$, and let $V \in T_z D_1$ be given. Let V_j be the complex components of V, so that

$$V(a_1)(z) = 2\,\mathrm{Re}\left[\sum V_j \left(\frac{\partial}{\partial z_j} a_1\right)(z)\right] = 2\,\mathrm{Re}\left(\sum V_j \beta_{1,j}(z)\right). \tag{6.17}$$

Set $W = d\psi_z(V) \in T_{\psi(z)} D_2$, so that

$$V(\dot{a}_2 \circ \psi)(z) = 2\,\mathrm{Re}\left(\sum W_j \dot{\beta}_{2,j}(z)\right). \tag{6.18}$$

It is very easy to derive (6.14) from (6.17), (6.18), (6.16), and the requirement that ϕ be a ν-symplectomorphism. For this it is helpful to identify V and W with tangent vectors to $T^*\mathbf{C}^n$, the second half of whose co-ordinates all vanish, and to notice that these tangent vectors to $T^*\mathbf{C}^n$ correspond under $d\phi$, by definitions.

7. RIEMANN MAPPINGS AND HCMA, PART 2

In this section we are going to show that if $\rho : B_n \to \mathbf{C}^n$ satisfies (1.1)–(1.4) and ρ extends to a bilipschitz map on $\overline{B}_n$ that is real analytic on $\overline{B}_n \setminus \{0\}$, then we can associate to ρ a holomorphic γ-symplectomorphism from an open set in $\mathcal{C}$ into $\mathcal{C}$ that also preserves Γ. The results of the preceeding section imply that this mapping induces a (nonlinear) transformation on functions that preserves HCMA, and we shall use the results of [L2] to show that the domain of this transformation contains a large collection of solutions of HCMA.

We should first point out why there exist mappings ρ with the properties just mentioned. If D is any strongly convex real-analytic domain in $\mathbf{C}^n$ which contains the origin, then there is a ρ with the above properties such that $\rho(B_n) = D$. This is just the real analytic version of Section 5. From [L1] it follows that the Kobayashi indicatrix I of D has real-analytic boundary, and that $\Psi : \overline{I} \to \overline{D}$ given by (5.1) is real analytic away from 0. We can take ρ as in Theorem 5.2, with σ required to be real analytic away from the origin. The existence of such a σ can be obtained by the method of the proof of Theorem 3.4.

The holomorphic symplectomorphism associated to ρ will be produced in two steps. We first define a certain lifting $\hat{\rho}$ of ρ, which will also be useful later on, and then we take a holomorphic extension of this lifting. The first step does not require real-analyticity.

Suppose that $\rho : B_n \to \mathbf{C}^n$ satisfies (1.1)–(1.4) and is C^2 on $B_n \setminus \{0\}$. Let $\widehat{B}_n$ denote the graph of ∂F_0 over B_n, so that

$$\widehat{B}_n = \{(z, \zeta) \in \mathcal{C} : \zeta = \bar{z},\ z \in B_n\}.$$

We define $\hat{\rho} : \widehat{B}_n \to \mathcal{C}$ as follows. Set $D = \rho(B_n)$ and $F = F_0 \circ \rho^{-1}$, and let $\Pi : \mathcal{C} \to \mathbf{C}^n$ denote the projection onto the first set of co-ordinates (i.e., the z part). We take

$$\hat{\rho} = (\rho \circ \Pi,\ \partial F \circ \rho \circ \Pi). \tag{7.1}$$

Thus the image of $\hat{\rho}$ is the graph $\widehat{D}$ of ∂F over D. Notice that F is C^1 across the origin even though ρ may not be; the derivatives of F at 0 all vanish.

The key properties of $\hat{\rho}$ are

$$\delta\hat{\rho}(\Gamma) = \delta i_{\widehat{B}_n}(\Gamma),\ \delta\hat{\rho}(\gamma) = \delta i_{\widehat{B}_n}(\gamma) \tag{7.2}$$

away from the origin $(0,0)$ in $\mathcal{C}$, where $i_{\widehat{B}_n} : \widehat{B}_n \to \mathcal{C}$ is the usual inclusion map. Indeed, under the map $\tau : D \to \widehat{D}$ defined by $\tau(z) = (z, \partial F(z))$ we have

$$\delta\tau(\Gamma) = \partial F, \tag{7.3}$$

by (6.5). The first equality in (7.2) follows easily from this observation, its counterpart for B_n and F_0, and the fact that $\delta\rho(\partial F) = \partial F_0$, by Theorem 2.2. The second equality follows from the first and the well-known fact that d commutes with the pull-back.

Suppose now that ρ is real analytic away from the origin. This implies that ρ has a holomorphic extension to a neighborhood of $\widehat{B}_n \setminus \{(0,0)\}$ in $\mathcal{C}$, because $\widehat{B}_n$ is totally real. We want to have control on the size of this neighborhood, among other things, and so we have to be careful.

We need a little more notation. Given $p \in \mathcal{C}$, let $|p|$ denote its Euclidean distance to $(0,0)$. For a subset E of $\mathcal{C}$ and a positive number ϵ we set

$$E_\epsilon = \{p \in \mathcal{C} : \operatorname{dist}(p, E) < \epsilon |p|\}. \tag{7.4}$$

Thus E_ϵ is a neighborhood of $E \setminus \{(0,0)\}$ that gets smaller as you proceed towards $(0,0)$.

THEOREM 7.5. *Suppose that* $\rho : B_n \to \mathbf{C}^n$ *satisfies* (1.1)–(1.4) *and that* ρ *extends to a bilipschitz mapping on* $\overline{B}_n$ *that is real-analytic on* $\overline{B}_n \setminus \{0\}$. *Set* $D = \rho(B_n)$. *Then there exist* ϵ, $\delta > 0$ *and* $C > 0$ *such that* $\hat{\rho}$ *has a holomorphic extension* $\phi : \widehat{B}_{n,\epsilon} \to \mathcal{C}$ *with the following properties:*

$$\phi(\widehat{B}_{n,\epsilon}) \supseteq \widehat{D}_\delta; \tag{7.6}$$

(7.7) ϕ *is a biholomorphism onto its image;*

$$\delta\phi(\Gamma) = \Gamma, \quad \delta\phi(\gamma) = \gamma; \tag{7.8}$$

$$C^{-1}|p| \le |\phi(p)| \le C|p| \text{ for } p \in \widehat{B}_{n,\epsilon}; \tag{7.9}$$

$$|d\phi| \le C, \ |(d\phi)^{-1}| \le C, \ \phi \text{ is bilipschitz on } \widehat{B}_{n,\epsilon}, \tag{7.10}$$
$$\text{and } |\nabla^2\phi(p)| \le C|p|^{-1} \text{ for } p \in \widehat{B}_{n,\epsilon}.$$

When we say that ρ is real analytic on $\overline{B}_n \setminus \{0\}$ we mean that it extends to be real analytic on some neighborhood of $\overline{B}_n \setminus \{0\}$. In particular F has an extension to a neighborhood of $\overline{D}$ that is real analytic except at the origin.

From Theorem 6.10 we obtain that ϕ induces a nonlinear transformation on functions that preserves HCMA. Before proving the theorem let us take a closer look at that transformation. For the rest of this section we assume that ρ, ϕ, and D are as in the theorem.

Let f be a real-valued C^2 function defined on some domain in $\mathbf{C}^n$, and let $N(f)$ denote the graph of ∂f in $\mathcal{C}$, i.e.,

$$N(f) = \{(z, \zeta) \in \mathcal{C} : \zeta_j = \partial_j f(z), \ z \text{ lies in the domain of } f\}.$$

Thus $N(f)$ is ν-Lagrangian, and if $N(f) \subseteq \widehat{B}_{n,\epsilon}$, then $\phi(N(f))$ is well defined and ν-Lagrangian. The following lemma provides a simple criterion for $N(f)$ to be contained in $\widehat{B}_{n,\epsilon}$ and $\phi(N(f))$ to be a graph, which is that f be close enough to F_0 in the appropriate way.

LEMMA 7.11. *Suppose that f is defined and C^2 on $B_n \setminus \{0\}$, that f and ∇f extend continuously across the origin, that $\nabla f(0) = 0$, and that*

$$|\nabla^2 F_0 - \nabla^2 f| \leq \beta \qquad \text{on } B_n \setminus \{0\} \tag{7.12}$$

for a sufficiently small β. Then $N(f) \subseteq \widehat{B}_{n,\epsilon}$ and $\phi(N(f)) = N(g)$ for some real-valued function g whose domain is almost $D \setminus \{0\}$ (in the sense described below). Moreover, g and ∇g extend continuously across the origin, $\nabla g(0) = 0$, and

$$|\nabla^2 F - \nabla^2 g| \leq C\beta \tag{7.13}$$

on the domain of g, for some $C > 0$ that does not depend on f. Furthermore, the normalizations $f(0) = 0$, $g(0) = 0$ force f and g to be positive away from the origin.

The domain of g is close to $D \setminus \{0\}$ in the sense that it is contained in a small neighborhood of $\overline{D}$, and that it contains $D \setminus \{0\}$ except for a small neighborhood of ∂D. In both cases "small" is controlled by β (linearly).

We restricted ourselves in this lemma to functions f defined on $B_n \setminus \{0\}$ merely for simplicity of exposition. We could just as well work with functions defined on a subdomain of $B_n \setminus \{0\}$, or whose domain lies in a sufficiently small neighborhood of $\overline{B}_n \setminus \{0\}$. The origin is excluded from the domain of f mostly for notational convenience; otherwise we could adjoin $(0,0)$ to $\widehat{B}_{n,\epsilon}$.

The correspondence between f and g can easily be reversed: if g is close enough to F then $\phi^{-1}(N(g))$ is a ν-Lagrangian graph, etc.

It is not very difficult to derive Lemma 7.11 from Theorem 7.5, and we shall do this after proving the latter.

Theorem 6.10 and the remarks thereafter imply that $\log g$ satisfies HCMA if $\log f$ does. It is not obvious that there exists an abundance of functions f for which the hypotheses of Lemma 7.11 hold and HCMA is satisfied by $\log f$. Fortunately this is true, because of the results of [L2]. There Lempert proves that if you are given a real-valued real-analytic function on ∂B_n which is sufficiently small in a certain sense (see [L2]), then you can find an extension h to $B_n \setminus \{0\}$ that is real-valued, real-analytic, and satisfies

$$\begin{cases} h & \text{is plurisubharmonic,} \\ h & \text{satisfies HCMA, and} \\ h(z) - \log|z| & \text{is bounded} \end{cases} \tag{7.14}$$

on $B_n \setminus \{0\}$. Lempert also obtains strong information about the behavior of h at 0, and about the dependence of h on the boundary data. Using his results it is not hard to show that $f = e^{2h}$ satisfies the hypotheses of Lemma 7.11 when the boundary data is small enough.

If f arises in this manner then $\frac{1}{2}\log g$ satisfies (7.14) on the domain of g. Indeed, we already know that $\log g$ satisfies HCMA, and $\frac{1}{2}\log g(z) - \log|z|$ is bounded because (7.13) implies that $|F(w) - g(w)| \leq C\beta|w|^2$ on the domain of

g. The plurisubharmonicity of g is more subtle, because it is not clear how to encode it into the behavior of Γ or γ on $N(g)$. In our case we can derive it from (7.13) if β is small enough, as follows. Lemma 2.9 implies that $\frac{\partial^2}{\partial z_j \partial \bar{z}_k} F(z)$ is a positive-definite matrix on $D \setminus \{0\}$, and using $\delta\rho(\partial\bar{\partial}F) = \partial\bar{\partial}F_0$ we can bound its eigenvalues away from 0. This is still true for the real-analytic extension of F to a neighborhood of $\overline{D} \setminus \{0\}$. Hence g is strictly plurisubharmonic if β is small enough, because of (7.13). This forces the Hessian of $\log g$ to have at most one nonpositive eigenvalue, which must be zero, since $\log g$ satisfies HCMA. Thus $\log g$ is plurisubharmonic, as desired.

Alternatively, we could have derived the strict plurisubharmonicity of g from an argument based on the minimum principle, as in the proof of Lemma 2.9. This approach has the advantage of being better suited to greater generality.

The rest of the section will be devoted to proving Theorem 7.5 and Lemma 7.11, starting with the former. To do this we must get some control on holomorphic extensions of ρ and F to open sets in the complexification of $\mathbf{C}^n$.

Let $\widetilde{\mathbf{C}}^n$ denote the complexification of $\mathbf{C}^n$, with $\mathbf{C}^n$ viewed as a real vector space. We may as well take $\widetilde{\mathbf{C}}^n$ to be $\mathbf{C}^n \times \mathbf{C}^n$, with the original $\mathbf{C}^n$ identified with $\{(z, \bar{z}) : z \in \mathbf{C}^n\}$ inside $\widetilde{\mathbf{C}}^n$.

Given $A \subseteq \mathbf{C}^n$ and $\epsilon > 0$, define

$$\widetilde{A}_\epsilon = \{w \in \widetilde{\mathbf{C}}^n : \operatorname{dist}(w, A) < \epsilon|w|\},$$

where $|w|$ denotes the Euclidean distance of w to the origin $(0,0)$ in $\widetilde{\mathbf{C}}^n$. This should not be confused with the similarly defined operation (7.4) on subsets of $\mathcal{C}$; generally instances of the latter come with $\hat{}$'s instead of $\tilde{}$'s, e.g., $\widehat{B}_{n,\epsilon}$ instead of $\widetilde{B}_{n,\epsilon}$.

Lemma 7.15. *There exist a, $C > 0$ so that ρ admits a bilipschitz extension $\tilde{\rho} : \widetilde{B}_{n,a} \to \widetilde{\mathbf{C}}^n$ that is holomorphic with respect to the complex structure on $\widetilde{\mathbf{C}}^n$ and which satisfies*

$$C^{-1}|w| \leq |\tilde{\rho}(w)| \leq C|w|, \tag{7.16}$$

$$|d\tilde{\rho}_w| + |w|\,|\nabla^2\tilde{\rho}(w)| + |w|^2\,|\nabla^3\tilde{\rho}(w)| \leq C, \tag{7.17}$$

$$|(d\rho_w)^{-1}| \leq C \tag{7.18}$$

for all $w \in \widetilde{B}_{n,a}$, and also

$$\text{for every } \alpha > 0 \text{ there is a } \beta > 0 \text{ so that } \tilde{\rho}(\widetilde{B}_{n,\alpha}) \supseteq \widetilde{D}_\beta. \tag{7.19}$$

We first show that there are a, $C > 0$ so that ρ admits a holomorphic extension $\tilde{\rho} : \widetilde{B}_{n,a} \to \widetilde{\mathbf{C}}^n$ such that

$$|\tilde{\rho}(w)| \leq C|w|. \tag{7.20}$$

We shall then check that this extension $\tilde{\rho}$ satisfies the other properties, at least if we shrink a and let C be larger.

Because ρ is real analytic on $\overline{B}_n \setminus \{0\}$, we can extend it to a holomorphic map of a neighborhood of $\overline{B}_n \setminus \{0\}$ in $\widetilde{\mathbf{C}}^n$ to $\widetilde{\mathbf{C}}^n$. In order to control the size of this neighborhood as you approach the origin we use the fact that ρ satisfies (1.3), which gives us the Cauchy integral representation

$$\rho(\lambda z) = \frac{1}{2\pi i} \int_{|\alpha|=1} \rho(\alpha z) \frac{\lambda}{(\alpha - \lambda)\alpha} d\alpha$$

for $z \in \overline{B}_n$, $\lambda \in \Delta$. [See (4.7).] Using this it is not hard to show that ρ can be extended holomorphically to $\tilde{\rho} : \widetilde{B}_{n,a} \to \widetilde{\mathbf{C}}^n$ for some $a > 0$, in such a way that (7.20) holds.

Using the Cauchy formula for derivatives it is easy to derive (7.17) with a slightly smaller value of a from (7.20). This implies in particular that $\tilde{\rho}$ is Lipschitz on $\widetilde{B}_{n,a}$. Hence the first inequality in (7.16) holds on $\widetilde{B}_{n,a}$ if a is shrunk sufficiently, since we already know that $|z| \leq C|\rho(z)|$ for $z \in \overline{B}_n$.

If we can show that (7.18) holds for $w \in \overline{B}_n \setminus \{0\}$, then a similar estimate holds on $\widetilde{B}_{n,a}$ for a small enough because of (7.17). The bound on $(d\tilde{\rho}_w)^{-1}$ for $w \in \overline{B}_n \setminus \{0\}$ follows directly from the corresponding bound for $(d\rho_w)^{-1}$, the fact that $\mathbf{C}^n$ is totally real inside $\widetilde{\mathbf{C}}^n$, and the complex linearity of $d\tilde{\rho}_w$.

Now let's check that $\tilde{\rho}$ is bilipschitz on $\widetilde{B}_{n,a}$ if we take a to be sufficiently small. We first observe that there is an $\eta > 0$ so that if w_1, $w_2 \in \widetilde{B}_{n,a}$ and $|w_1 - w_2| \leq \eta |w_1|$, then

$$|\tilde{\rho}(w_1) - \tilde{\rho}(w_2)| \geq C^{-1}|w_1 - w_2|. \tag{7.21}$$

This uses (7.17), (7.18), Taylor's theorem, and our right to shrink a. Fix such an η. If a is small enough, and if w_1, $w_2 \in \widetilde{B}_{n,a}$ satisfy $|w_1 - w_2| \geq \eta(\max(|w_1|, |w_2|))$, then (7.21) also holds. To see this we choose z_1, $z_2 \in \overline{B}_n \setminus \{0\}$ such that $|w_i - z_i| < a|w_i|$, $i = 1, 2$, and use the bilipschitzness of ρ to get

$$\begin{aligned} C^{-1}|z_1 - z_2| &\leq |\rho(z_1) - \rho(z_2)| \\ &\leq |\tilde{\rho}(w_1) - \tilde{\rho}(w_2)| + |\tilde{\rho}(z_1) - \tilde{\rho}(w_1)| + |\tilde{\rho}(z_2) - \tilde{\rho}(w_2)| \\ &\leq |\tilde{\rho}(w_1) - \tilde{\rho}(w_2)| + Ca(|w_1| + |w_2|) \\ &\leq |\tilde{\rho}(w_1) - \tilde{\rho}(w_2)| + Ca\eta^{-1}|w_1 - w_2|. \end{aligned}$$

On the other hand

$$\begin{aligned} |z_1 - z_2| &\geq |w_1 - w_2| - |z_1 - w_1| - |z_2 - w_2| \\ &\geq |w_1 - w_2| - a(|w_1| + |w_2|) \\ &\geq |w_1 - w_2| - 2a\eta^{-1}|w_1 - w_2|. \end{aligned}$$

If a is small enough, these two estimates combined to give (7.21). Thus $\tilde{\rho}$ is bilipschitz on $\widetilde{B}_{n,a}$ if a is small enough.

There are many ways that (7.19) can be proved. One of them uses the inverse function theorem. This completes the proof of Theorem 7.15.

LEMMA 7.22. *There exist $b, C > 0$ so that F admits a holomorphic extension $\widetilde{F} : \widetilde{D}_b \to \mathbf{C}$ such that $|\widetilde{F}(w)| \leq C|w|^2$, $|\nabla\widetilde{F}(w)| \leq C|w|$, $|\nabla^2\widetilde{F}(w)| \leq C$, and $|\nabla^3\widetilde{F}(w)| \leq C|w|^{-1}$ on $\widetilde{D}_b$.*

We can in fact express $\widetilde{F}$ explicitly in terms of $\tilde{\rho}$, because F_0 has an explicit holomorphic extension $\widetilde{F}_0$ to $\widetilde{\mathbf{C}}^n$. Indeed,

$$\widetilde{F}_0((z, z')) = \sum z_j z_j'$$

is holomorphic on $\widetilde{\mathbf{C}}^n$, and $\widetilde{F}_0((z, \bar{z})) = F_0(z)$. Thus we can take $\widetilde{F} = \widetilde{F}_0 \circ \tilde{\rho}^{-1}$, and it is easy to see that this works.

There is still a little more that we need to do to prove Theorem 7.5. Let $\tau_0 : B_n \to \widehat{B}_n$ and $\tau : D \to \widehat{D}$ be defined in the obvious way, that is, $\tau_0(z) = (z, \bar{z})$, $\tau(z) = (z, \partial F(z))$. We need to have holomorphic extensions $\tilde{\tau}_0$, $\tilde{\tau}$ of τ_0, τ from domains in $\widetilde{\mathbf{C}}^n$ into $\mathcal{C}$.

For τ_0 this is a matter of chasing definitions. Using our identifications of $\widetilde{\mathbf{C}}^n$ and $\mathcal{C}$ with $\mathbf{C}^n \times \mathbf{C}^n$, and our identification of $\mathbf{C}^n$ with $\{(z, z') \in \widetilde{\mathbf{C}}^n : z' = \bar{z}\}$, we can simply define $\tilde{\tau}_0$ by $\tilde{\tau}_0((z, z')) = (z, z')$, i.e., $\tilde{\tau}_0$ is the identity.

The story for $\tilde{\tau}$ is a little more complicated. There is a holomorphic extension $\tilde{\tau} : \widetilde{D}_b \to \mathcal{C}$ of τ which can be given in terms of $\widetilde{F}$. One can check that

$$|w|^{-1}|\tilde{\tau}(w)| + |\nabla\tilde{\tau}(w)| + |w|\,|\nabla^2\tilde{\tau}(w)| \leq C \tag{7.23}$$

for all $z \in \widetilde{D}_b$, using the corresponding estimates for $\widetilde{F}$.

A key fact about $\tilde{\tau}$ is that

$$|(d\tilde{\tau}_w)^{-1}| \leq C \tag{7.24}$$

for $w \in \widetilde{D}_e$ if $e > 0$ is small enough. Because of the estimate on $\nabla^2\tilde{\tau}$ in (7.23) it suffices to check this on $D \setminus \{0\}$.

We should explain why $d\tilde{\tau}_w$ is invertible for $w \in D\setminus\{0\}$. This is a consequence of the following two observations: $d\tau_w$ is an injective real-linear map of $T_w D$ into $T_{\tau(w)}\mathcal{C}$ whose range is $T_{\tau(w)}\widehat{D}$, and $T_{\tau(w)}\widehat{D}$ is a totally real subspace of $T_{\tau(w)}\mathcal{C}$ whose complexification is all of $T_{\tau(w)}\mathcal{C}$. The latter reduces to the nondegeneracy of $\partial\overline{\partial}F$. The estimate (7.24) follows from the quantitative versions of these two facts. In the case of the second, the quantitative version comes from the bounds $|\partial\overline{\partial}F| \leq C$, $|(\partial\overline{\partial}F)^n| \geq C^{-1}$, which can be derived from $\delta\rho(\partial\overline{\partial}F) = \partial\overline{\partial}F_0$ and the bilipschitzness of ρ.

There are two more pieces of information about $\tilde{\tau}$ that we require. The first is that there is an $e > 0$ such that $\tilde{\tau}$ is bilipschitz on $\widetilde{D}_e$. This can be derived from (7.23), (7.24), and the bilipschitzness of τ on $\overline{D}$, just like the proof of the corresponding result for $\tilde{\rho}$. The second is that for each $\epsilon > 0$ there is a $\delta > 0$ so that $\tilde{\tau}(\widetilde{D}_\epsilon) \supseteq \widehat{D}_\delta$, which can be obtained from the inverse function theorem, for instance.

With all of this information on $\tilde{\rho}$, $\tilde{\tau}_0$, $\tilde{\tau}$ it is easy to prove Theorem 7.5. We define ϕ by $\phi = \tilde{\tau} \circ \tilde{\rho} \circ \tilde{\tau}_0^{-1}$. It is easy to see that there is an $e > 0$ so that ϕ

is defined and holomorphic on $\widehat{B}_{n,\epsilon}$ and that ϕ satisfies (7.6), (7.7), (7.9), and (7.10). This leaves (7.8), which follows from (7.2), since $\phi\mid_{\widehat{B}_n} = \hat{\rho}$, $\widehat{B}_n$ is totally real, and γ, Γ are holomorphic.

Let us prove Lemma 7.11. Clearly (7.12) implies

$$|\nabla F_0(w) - \nabla f(w)| \leq C\beta |w| \qquad \text{on } B_n, \tag{7.25}$$

and hence $N(f) \subseteq \widehat{B}_{n,\epsilon}$ if β is small enough. Thus $\phi(N(f))$ is well defined, and we want to show that it is a graph, and even a Lipschitz graph, if β is sufficiently small.

This is the same as showing that Π restricted to $\phi(N(f))$ is one-to-one and even bilipschitz, or, equivalently, that $\Pi \circ \phi \circ \tau_f$ is bilipschitz, where $\tau_f : B_n \to N(f)$ is given by $\tau(z) = (z, \partial f(z))$. To prove this we use the fact that $\Pi \circ \phi \circ \tau_0 = \rho$ is bilipschitz and that f is close to F_0. Using (7.10), (7.12), and (7.25) we get that

$$|\nabla(\phi \circ \tau_0 - \phi \circ \tau_f)| \leq C\beta \tag{7.26}$$

on $B_n \setminus \{0\}$, and so $\phi \circ \tau_0 - \phi \circ \tau_f$ is Lipschitz on $\overline{B}_n$ with norm $\leq C\beta$. Hence $\rho - \Pi \circ \phi \circ \tau_f$ has Lipschitz norm $\leq C\beta$, and $\Pi \circ \phi \circ \tau_f$ is bilipschitz if β is small enough. This implies that $\phi(N(f))$ is the graph of a Lipschitz function over the domain $D_1 = \Pi(\phi(N(f)))$. It is not hard to see that D_1 must be close to D, in the sense explained just after the statement of Lemma 7.11.

This Lipschitz function on D_1 must also be C^1 away from the origin. because f is C^2 on $B_n \setminus \{0\}$. Since $\phi(N(f))$ is ν-Lagrangian we conclude that $\phi(N(f)) = N(g)$ for some real-valued function g on D_1 that is C^2 on $D_1 \setminus \{0\}$, and g also satisfies $|\nabla^2 g| \leq C$ on $D_1 \setminus \{0\}$, $\nabla g(0) = 0$.

Let us check (7.13). Let $\Pi' : \mathcal{C} \to \mathbf{C}^n$ be the obvious projection onto the second set of co-ordinates on $\mathcal{C}$. By definitions we have

$$(\partial F) \circ \Pi \circ \phi \circ \tau_0 = \Pi' \circ \phi \circ \tau_0,$$
$$(\partial g) \circ \Pi \circ \phi \circ \tau_f = \Pi' \circ \phi \circ \tau_f.$$

Using this and (7.26) we conclude that

$$|\nabla\left((\partial g) \circ \Pi \circ \phi \circ \tau_f - (\partial F) \circ \Pi \circ \phi \circ \tau_0\right)| \leq C\beta.$$

If we can show that

$$|\nabla\left((\partial F) \circ \Pi \circ \phi \circ \tau_0 - (\partial F) \circ \Pi \circ \phi \circ \tau_f\right)| \leq C\beta, \tag{7.27}$$

then we get

$$|\nabla\left((\partial g) \circ \Pi \circ \phi \circ \tau_f - (\partial F) \circ \Pi \circ \phi \circ \tau_f\right)| \leq C\beta,$$

which implies (7.13), because $\Pi \circ \phi \circ \tau_f$ is bilipschitz, as we have seen.

It is not hard to verify (7.27) using (7.26),

$$|\phi(\tau_0(z)) - \phi(\tau_f(z))| \leq C\beta|z|, \text{ and} \tag{7.28}$$

$$|\nabla^2 F| \leq C, \ |\nabla^3 F(z)| \leq C|z|^{-1} \text{ on a neighborhood of } \overline{D} \setminus \{0\}. \tag{7.29}$$

Clearly (7.28) follows from (7.25) and (7.10), while (7.29) follows from Lemma 7.22. [We could prove (7.29) more directly, using the equation $F = F_0 \circ \rho^{-1}$ to reduce to the corresponding inequalities on $\nabla^j \rho$, and then using a Cauchy integral representation of ρ to get the appropriate estimates on $\nabla^j \rho$.]

The last statement in Lemma 7.11, that the normalizations $f(0) = 0$, $g(0) = 0$ force f, g to be positive away from the origin, follows from $F_0(z) = |z|^2$, $F(z) \geq C^{-1}|z|^2$, and

$$|f(z) - F_0(z)| \leq C\beta|z|^2, \ |g(z) - F(z)| \leq C\beta|z|^2.$$

The last two inequalities come from (7.12), (7.13), $f(0) = \nabla f(0) = 0$, and $g(0) = \nabla g(0) = 0$. This completes the proof of Lemma 7.11.

8. RIEMANN MAPPINGS AND LIFTINGS TO $\mathcal{C}$

In the previous section we saw that if $\rho : B_n \to \mathbf{C}^n$ satisfies (1.1)–(1.4) and is C^2 away from the origin then it has a lifting $\hat{\rho} : \widehat{B}_n \to \mathcal{C}$ that satisfies (7.2). In this section we record a converse.

PROPOSITION 8.1. *Suppose that $\psi : \widehat{B}_n \to \mathcal{C}$ satisfies the following properties: ψ is C^1 away from $(0,0)$ and continuous across $(0,0)$; ψ maps $(0,0)$ to $(0,0)$; $\delta\psi(\Gamma) = \delta i_{\widehat{B}_n}(\Gamma)$ away from $(0,0)$; and $\rho = \Pi \circ \psi \circ \tau_0$ satisfies (1.1) and (1.2), where $\tau_0 : B_n \to \widehat{B}_n$ is given by $\tau_0(z) = (z, \overline{z})$, as usual. Then ρ also satisfies (1.3) and (1.4), and $\psi = \hat{\rho}$, where $\hat{\rho}$ as defined in (7.1).*

Set $D = \rho(B_n)$. There is a continuous $(1,0)$-form $\beta = \sum \beta_j dz_j$ on D such that $\psi = (\rho, \{\beta_j \circ \rho\}_{j=1}^n)$, and β is C^1 away from the origin. Set $G = \psi(\widehat{B}_n)$, the graph of β.

Define $\tau : D \to G$ by (6.4). Thus $\psi \circ \tau_0 = \tau \circ \rho$, and therefore

$$\delta(\psi \circ \tau_0)(\Gamma) = \delta\tau_0(\delta i_{\widehat{B}_n}(\Gamma)) = \delta\tau_0(\Gamma) = \partial F_0$$

equals

$$\delta(\tau \circ \rho)(\Gamma) = \delta\rho(\delta\tau(\Gamma)) = \delta\rho(\beta)$$

away from the origin. We have used (6.5), applied to τ_0 and τ, in the last of each of these chains of equalities. Hence $\delta(\rho^{-1})(\partial F_0) = \beta$ is a $(1,0)$ form, and Theorem 2.2 tells us that ρ satisfies (1.3) and (1.4), and that $\beta = \partial F$, $F = F_0 \circ \rho^{-1}$. This proves the proposition.

The condition

$$\delta\psi(\Gamma) = \delta i_{\widehat{B}_n}(\Gamma) \qquad \text{on } \widehat{B}_n \setminus \{(0,0)\} \tag{8.2}$$

admits an interesting reformulation when ψ is real-analytic on $\widehat{B}_n \setminus \{(0,0)\}$. In that case ψ has a holomorphic extension ϕ to a connected open set containing $\widehat{B}_n \setminus \{(0,0)\}$, and (8.2) is equivalent to

$$\delta\phi(\Gamma) = \Gamma, \tag{8.3}$$

since $\widehat{B}_n$ is totally real. This is also equivalent to

$$\delta\phi(\gamma) = \gamma, \; d\phi(V) = V, \; V = \sum \zeta_j \frac{\partial}{\partial \zeta_j}. \tag{8.4}$$

Let's check this. Clearly (8.3) implies that $\delta\phi(\gamma) = \gamma$, because the pull-back commutes with d. If we assume that $\delta\phi(\gamma) = \gamma$, then it is easy to see that $\delta\phi(\Gamma) = \Gamma$ iff $d\phi(V) = V$, because Γ and $i(V)\gamma$ differ only by a multiplicative constant. These two observations imply the equivalence between (8.3) and (8.4).

In order to obtain a reasonable parameterization of a space of ψ's that satisfy the hypotheses of Proposition 8.1 using implicit function theorem methods it is probably better to work with (8.4) instead of (8.2). This is because $\delta\phi(\gamma) = \gamma$ is a well-behaved nondegenerate constraint on ϕ (as in Section 4 of [EM]), and $d\phi(V) = V$ is linear in ϕ. As it stands, however, this reformulation only works for real-analytic ψ's, and it would be preferable to deal with spaces of ψ's that are merely pretty smooth.

There is a simple way in which these observations permit us to generate lots of ψ's that satisfy the hypotheses of Proposition 8.1, in a more painless fashion than by using the implicit function theorem, which is to exponentiate vector fields.

Fix $\psi_0 : \widehat{B}_n \to \mathcal{C}$ which satisfies the hypotheses of the proposition. Set $E = \psi_0(\widehat{B}_n)$, and let E_ϵ be as in (7.4). Suppose that W is a holomorphic vector field defined on $E_{2\epsilon}$ for some $\epsilon > 0$, and that

$$|W(p)| \leq C|p| \qquad \text{for } p \in E_{2\epsilon}. \tag{8.5}$$

Then there exists a $t_0 > 0$ so that we can find functions $g_t : E_\epsilon \to E_{2\epsilon}$ for $-t_0 \leq t \leq t_0$ such that $g_0 = id$,

$$\dot{g}_t = W \circ g_t,$$

and such that each g_t is a bilipschitz biholomorphism onto its image that satisfies (7.9) on E_ϵ. If we also have $L_W\Gamma = 0$, then $\delta g_t(\Gamma) = \Gamma$ for each t, and $g_t \circ \psi_0$ also satisfies the hypotheses of Proposition 8.1.

Let us look more closely at the condition $L_W\Gamma = 0$. Analogous to (8.3) and (8.4), this is equivalent to $L_W\gamma = 0$, $L_W V = 0$, where V is the holomorphic vector field in (8.4). In particular W must be γ-symplectic.

If G is any complex-valued holomorphic function on $E_{2\epsilon}$ such that

$$|\nabla G(p)| \leq C|p| \qquad \text{on } E_{2\epsilon}, \tag{8.6}$$

then the symplectic gradient of G with respect to γ provides a holomorphic vector field W on $E_{2\epsilon}$ that satisfies (8.5) and $L_W\gamma = 0$. Let's compute what the stronger condition $L_W(\Gamma) = 0$ means in terms of G.

Using a well-known formula (see p. 70 of [Wa]) we have

$$L_W(\Gamma) = d(i(W)\Gamma) + i(W)d\Gamma = d(\Gamma(W)) - 2\sqrt{-1}i(W)\gamma = d(\Gamma(W)) - 2\sqrt{-1}dG, \tag{8.7}$$

since $i(W)\gamma = dG$ if W is the γ-symplectic gradient of G. On the other hand,

$$\Gamma(W) = 2\sqrt{-1}\gamma(W, V) = 2\sqrt{-1}V(G).$$

Plugging this into (8.7) we get that $L_W(\Gamma) = 0$ if $G - V(G)$ is constant. This is equivalent to the condition

$$G(z,\zeta) - \frac{\partial}{\partial\lambda}\Big|_{\lambda=1} G(z,\lambda\zeta) \qquad \text{is constant on } E_{2\epsilon}. \tag{8.8}$$

Thus if G is holomorphic on $E_{2\epsilon}$ and satisfies (8.6) and (8.8), then its γ-symplectic gradient W is a holomorphic vector field that satisfies (8.5) and $L_W\Gamma = 0$, so that flowing along W produces new ψ's for which the hypotheses of Proposition 8.1 hold.

Consider now the special case where $\psi_0 = id$, so that $E = \widehat{B}_n$. We want to give a simpler description of the set of admissible G's, one that will make it clearer how they can be produced.

First we impose a normalization on G. From (8.6) it follows that G extends continuously to $(0,0)$, and we may as well demand that G vanishes there. Then (8.8) becomes

$$G(z,\zeta) - \frac{\partial}{\partial\lambda}\Big|_{\lambda=1} G(z,\lambda\zeta) = 0, \tag{8.9}$$

since (8.6) ensures that the second term tends to zero as (z,ζ) approaches $(0,0)$.

Let us show that the set

(8.10) $\{$complex-valued functions H on $\overline{B}_n$: there is a $\delta > 0$ and a holomorphic function G on $\widehat{B}_{n,\delta}$ such that G satisfies (8.6), (8.8), $G(0,0) = 0$, and $H(z) = G(z,\overline{z})\}$

is the same as

(8.11) $\{$complex-valued functions H on $\overline{B}_n$: H is continuous on $\overline{B}_n$, real-analytic on ∂B_n, $H(0) = 0$, $|H(z)| \le C|z|^2$ for some $C > 0$, and $\overline{\lambda}^{-1} H(\lambda z)$ is holomorphic in $\lambda \in \Delta$ for each $z \in \partial B_n\}$.

It is pretty easy to see that (8.10) is contained in (8.11). This is perhaps made clearer by the observation that if G is holomorphic on a neighborhood of $\widehat{B}_n \setminus \{(0,0)\}$, and if $\check{G}$ is defined on B_n by $\check{G}(z) = G(z,\overline{z})$, then (8.9) implies that

$$\check{G}(z) - \frac{\partial}{\partial\overline{\lambda}}\Big|_{\lambda=1} \check{G}(\lambda z) = 0 \tag{8.12}$$

for all $z \in B_n \setminus \{0\}$.

The reverse inclusion is not difficult either. We first observe that H admits the Cauchy integral representation

$$\begin{aligned} H(\lambda z) &= \frac{\overline{\lambda}}{2\pi i} \int_{|\alpha|=1} \left(H(\alpha z)\overline{\alpha}^{-1}\right) \left(\frac{1}{\alpha - \lambda} - \frac{1}{\alpha}\right) d\alpha \\ &= \frac{\overline{\lambda}}{2\pi i} \int_{|\alpha|=1} H(\alpha z)\frac{\lambda}{(\alpha - \lambda)} d\alpha \end{aligned} \tag{8.13}$$

for all $\lambda \in \Delta$, $z \in \partial B_n$ when H lies in (8.11). This formula and the real-analyticity of H on ∂B_n imply that H is real-analytic on $\overline{B}_n \setminus \{0\}$, and that there is a $\delta > 0$ and a holomorphic function G on $\widehat{B}_{n,\delta}$ such that $H(z) = G(z, \overline{z})$, $G(0,0) = 0$, and G satisfies (8.6). The fact that $\overline{\lambda}^{-1} H(\lambda z)$ is holomorphic in λ implies that H satisfies (8.12), from which it follows that G satisfies (8.9) on $\widehat{B}_n \setminus \{(0,0)\}$, and hence on $\widehat{B}_{n,\delta}$, since $\widehat{B}_n$ is totally real. This proves that (8.11) is contained in (8.10), and so they are equal.

It is not hard to show that if $K(z)$ is any complex-valued real-analytic function on ∂B_n, then there is an H in (8.11) so that

$$H(\lambda z) = \frac{\overline{\lambda}}{2\pi i} \int_{|\alpha|=1} K(\alpha z) \frac{\lambda}{(\alpha - \lambda)} d\alpha \tag{8.14}$$

for all $\lambda \in \Delta$, $z \in \partial B_n$. Indeed, we can take (8.14) to be the definition of H, after noticing that different choices of λ and z give the same value for H as long as λz is held fixed. The required properties of H follow from this definition and well-known facts about the Cauchy integral.

There is another way to describe and produce Riemann mappings in the real analytic category using Proposition 8.1, which is to use the method of generating functions. The relevance of this method in the present context was brought to the author's attention by Tom Duchamp. In the following discussion an outline of the key ideas will be presented without the technical details, which are fairly straightforward and quite similar to issues that we've already encountered in Section 7. The reader may wish to consult p. 135–136 of [BFG] for a brief description of the method of generating functions.

In view of Theorem 7.5 and Proposition 8.1 there is a correspondence between real-analytic Riemann mappings and biholomorphisms ϕ from an open subset of $\mathcal{C}$ onto another open subset of $\mathcal{C}$ such that $\delta\phi(\gamma) = \gamma$, $\delta\phi(\Gamma) = \Gamma$, and ϕ and its domain satisfy certain technical conditions (as in the statement of Theorem 7.5). Let $\tilde{\mathcal{C}}$ be another copy of $\mathcal{C}$, with global co-ordinates $\tilde{z}_j$, $\tilde{\zeta}_j$ and associated differential forms $\tilde{\gamma}$ and $\tilde{\Gamma}$, as in Section 6, and let us view ϕ as a map from a subset of $\mathcal{C}$ into $\tilde{\mathcal{C}}$. Set $M = \mathcal{C} \times \tilde{\mathcal{C}}$, and let $L = L(\phi)$ denote the submanifold of M that is the graph of ϕ.

The main properties of ϕ — that it be holomorphic and satisfy $\delta\phi(\tilde{\gamma}) = \gamma$, $\delta\phi(\tilde{\Gamma}) = \Gamma$ — can be reformulated in terms of L as the requirements that L be a holomorphic submanifold, with L lagrangian with respect to the holomorphic symplectic form $\gamma - \tilde{\gamma}$ on M, and with $\delta i_L(\Gamma - \tilde{\Gamma}) = 0$, where $i_L : L \to M$ is the obvious inclusion mapping.

Consider the holomorphic 1-form θ on M defined by $\theta = \sum(\zeta_j dz_j + \tilde{z}_j d\tilde{\zeta}_j)$. Clearly $d\theta = \gamma - \tilde{\gamma}$, and so if the domain of ϕ is simply connected (which is true in the cases that we are interested in), then there will be a holomorphic function f on L such that $df = \delta i_L(\theta)$.

Define a projection $P : M \to \mathbf{C}^n \times \mathbf{C}^n$ by $P((z, \zeta, \tilde{z}, \tilde{\zeta})) = (z, \tilde{\zeta})$. If ϕ is a small perturbation of the identity (in the bilipschitz topology), then P induces a biholomorphism from L onto an open subset U of $\mathbf{C}^n \times \mathbf{C}^n$. Define $\check{f}$ on U by

$\check{f} \circ P = f$. As in [BFG], the fact that $df = \delta i_L(\theta)$ is equivalent to

$$L = \left\{ (z,\zeta,\tilde{z},\tilde{\zeta}) \in M : (z,\tilde{\zeta}) \in U, \zeta_j = \frac{\partial \check{f}(z,\zeta)}{\partial z_j},\ \tilde{z}_j = \frac{\partial \check{f}(z,\tilde{\zeta})}{\partial \tilde{\zeta}_j} \right\}. \tag{8.15}$$

The simplest example of this is $\phi = \mathrm{id}$, $f(z,\tilde{\zeta}) = \sum z_j \tilde{\zeta}_j$.

Conversely, if you are given a holomorphic function $\check{f}$ on an open set $U \subseteq \mathbf{C}^n \times \mathbf{C}^n$ which is a small perturbation of $\sum z_j \tilde{\zeta}_j$ in the C^2 topology, then (8.15) defines a holomorphic Lagrangian submanifold of $(M, \gamma - \tilde{\gamma})$ that is the graph of a biholomorphism ϕ from an open set in $\mathcal{C}$ onto another one in $\tilde{\mathcal{C}}$ that satisfies $\delta\phi(\tilde{\gamma}) = \gamma$.

In our case we want to have also the condition $\delta\phi(\tilde{\Gamma}) = \Gamma$, or, equivalently, $\delta i_L(\Gamma - \tilde{\Gamma}) = 0$. Let us reformulate this in terms of f and $\check{f}$. Clearly

$$\theta - \Gamma + \tilde{\Gamma} = d(\sum \tilde{z}_j \tilde{\zeta}_j),$$

and so $\delta i_L(\Gamma - \tilde{\Gamma}) = 0$ if and only if

$$f = \left(\sum \tilde{z}_j \tilde{\zeta}_j\right) |_L + \text{ a constant},$$

or, equivalently,

$$\check{f}(z,\tilde{\zeta}) = \sum \frac{\partial \check{f}(z,\tilde{\zeta})}{\partial \tilde{\zeta}_j} \tilde{\zeta}_j + \text{ a constant},$$

by (8.15). This last is equivalent to

$$\check{f}(z,\tilde{\zeta}) - \frac{\partial}{\partial \lambda}\Big|_{\lambda=1} (\check{f}(z,\lambda\tilde{\zeta})) = \text{ a constant.} \tag{8.16}$$

The preceeding is really just a summary of the classical method of generating functions, with suitable adjustments to allow for the holomorphic context and the additional constraint $\delta\phi(\tilde{\Gamma}) = \Gamma$. This method provides a correspondence between biholomorphic mappings ϕ which satisfy $\delta\phi(\gamma) = \gamma$, $\delta\phi(\Gamma) = \Gamma$, and which are small perturbations of the identity, and holomorphic complex valued functions $\check{f}(z,\tilde{\zeta})$ defined on open subsets of $\mathbf{C}^n \times \mathbf{C}^n$ that satisfy (8.16) and which are small perturbations of $\sum z\tilde{\zeta}$.

In our setting there are some additional technical issues which should be addressed, and, as promised earlier, they will only be sketched. Let us begin by considering ϕ's that arise from real-analytic Riemann mappings as in Theorem 7.5, and which, furthermore, are small perturbations of the identity in the Lipschitz topology on their domains. We want to know what kinds of functions $\check{f}$ arise from the ϕ's.

To do this some additional notation will be helpful. Let $\mathcal{E}$ denote another copy of $\mathbf{C}^n \times \mathbf{C}^n$, but with co-ordinates $z_j, \tilde{\zeta}_j$. Think of $\mathcal{E}$ as being the copy of $\mathbf{C}^n \times \mathbf{C}^n$ on which $\check{f}$ lives. Let $X : \mathcal{C} \to \mathcal{E}$ be the obvious identification.

Let $\mathcal{F}$ denote the set of holomorphic complex-valued functions $\check{f}(z,\tilde{\zeta})$ defined on $X(\widehat{B}_{n,\epsilon})$ for some $\epsilon > 0$ such that

$$(8.17) \qquad |d\check{f}(z,\tilde{\zeta})| \leq C(|z|+|\zeta|),$$

$$(8.18) \qquad |\check{f}(z,\tilde{\zeta})| \leq C(|z|+|\zeta|)^2,$$

$$(8.19) \qquad \check{f}(z,\tilde{\zeta}) - \frac{\partial}{\partial\lambda}\bigg|_{\lambda=1}\left(\check{S}(z,\lambda\tilde{\zeta})\right) = 0, \text{ and}$$

$$(8.20) \qquad |d(\check{f}(z,\tilde{\zeta}) - \textstyle\sum z_j\tilde{\zeta}_j)|(|z|+|\tilde{\zeta}|)^{-1} \text{ is small}$$

for some $C > 0$ and all z, $\zeta \in X(\widehat{B}_{n,\epsilon})$. (Of course $\mathcal{F}$ is not really a set until you specify a choice of what "small" means.) Then every $\check{f} \in \mathcal{F}$ arises from a ϕ as above, and every $\check{f} \in \mathcal{F}$ gives rise to a ϕ as above, according to the correspondence described earlier. The task of checking this is straightforward, but there is one point that should be mentioned. In the correspondence between ϕ's and $\check{f}$'s discussed above, $\check{f}$ was determined only up to an additive constant. In our context it is natural to get rid of this ambiguity by imposing the normalizing assumption that $\check{f}(z,\tilde{\zeta})$ vanishes at $(0,0)$. This normalization gives rise to (8.18) and the simplification (8.19) of (8.16).

Notice that in passing from ϕ's to $\check{f}$'s and back to ϕ's you lose a little on the size of the domains and the extent to which the perturbations are small.

Notice also that the class $\mathcal{F}$ of $\check{f}$'s corresponds under the identification $X : \mathcal{C} \to \mathcal{E}$ to the class of functions in (8.10), and hence also (8.11). More precisely, if $\check{f} \in \mathcal{F}$, then $\check{f}(z,\tilde{\zeta}) - \sum z_j\tilde{\zeta}_j$ composed with X gives rise to a function in (8.10) which is small, and this procedure can be reversed.

Thus we have a way to parameterize the set of real-analytic Riemann mappings that are small perturbations of the identity by the class of functions $\mathcal{F}$. Similarly we can parameterize the set of real-analytic Riemann mappings which are small perturbations of some fixed real-analytic Riemann mapping ρ_1 by a nice class of functions as follows. Let ϕ_1 correspond to ρ_1 as in Theorem 7.5. Using Theorem 7.5 and Proposition 8.1 it is easy to use ϕ_1 to produce a one-to-one correspondence between real-analytic Riemann mappings that are small perturbations of ρ_1 and real-analytic Riemann mappings which are small perturbations of the identity. (This sort of invariance will be discussed more fully in Section 12.) Thus we can reduce to the case where ρ_1 is the identity, and we can again use $\mathcal{F}$ as our class of functions.

Alternatively, we can take the following approach. Let ρ be a real-analytic Riemann mapping that is a small perturbation of ρ_1, and let ϕ be as in Theorem 7.5. Then $\alpha = \phi \circ \phi_1^{-1}$ is a holomorphic mapping which is a small perturbation of the identity on $(\phi_1(\widehat{B}_n))_\eta \subseteq \mathcal{C}$ for some $\eta > 0$ (the notation in (7.4) is being employed here), and α also satisfies $\delta\alpha(\gamma) = \gamma$, $\delta\alpha(\Gamma) = \Gamma$. The earlier story about generating functions can also be applied in this situation. In this way we can again produce a correspondence between real-analytic Riemann mappings which are small perturbations of ρ_1 and a certain class of functions. This approach is obviously very similar to the previous one, but it is not quite the same.

These parameterizations of portions of the space of Riemann mappings are compatible with the complex structure on the space of Riemann mappings which was mentioned in the introduction and which is discussed in more detail in Section 10. In particular a holomorphic family of generating functions gives rise to a holomorphic family of Riemann mappings. From the perspective of infinite-dimensional manifolds, however, these parameterizations are not so great. This is due in large measure to the fact that the relevant topological vector spaces are pretty awful.

9. SPACES OF RIEMANN MAPPINGS, SPACES OF DOMAINS

We now shift our attention away from the properties of individual Riemann mappings and domains, and instead look at spaces of such objects and geometric structures on them.

For the purposes of this endeavor it will be helpful to impose additional smoothness conditions. Usually we don't need much extra regularity, but for reasons of exposition it seems preferable to work in the C^∞ category than to keep track of what is used when. There are occasions, cousins of Sections 7 and 8, when it is desirable to have real-analyticity, and so we give notation to that case as well.

Let $\mathcal{R}^\infty$ the space of maps $\rho : \overline{B}_n \to \mathbf{C}^n$ which satisfy (1.1), (1.3), (1.4), and the following strengthened version of (1.2):

$$\text{(9.1)} \qquad \rho \text{ is bilipschitz on } \overline{B}_n \text{ and } C^\infty \text{on } \overline{B}_n \setminus \{0\}.$$

We define $\mathcal{R}^\omega$ similarly, but with ρ required to be real-analytic on $\overline{B}_n \setminus \{0\}$. (That is, ρ extends to a real-analytic mapping on a neighborhood of $\overline{B}_n \setminus \{0\}$.)

Sometimes it is better to work not with $\mathcal{R}^\infty$ but with its counterpart in $\mathcal{C}$, as indicated in Sections 7 and 8. Let $\widehat{\mathcal{R}}^\infty$ denote the space of mappings ψ of the closure $\overline{\widehat{B}_n}$ of $\widehat{B}_n$ into $\mathcal{C}$ such that:

(9.2) $\psi((0,0)) = (0,0)$;

(9.3) ψ is a bilipschitz map of $\overline{\widehat{B}_n}$ into $\mathcal{C}$ that is C^∞ away from $(0,0)$;

(9.4) $\rho = \Pi \circ \psi \circ \tau_0$ satisfies (9.1), where $\tau_0 : B_n \to \overline{\widehat{B}_n}$ is given by $\tau_0(z) = (z, \overline{z})$;

(9.5) $\delta\psi(\Gamma) = \delta i_{\widehat{B}_n}(\Gamma)$ and $\delta\psi(\gamma) = \delta i_{\widehat{B}_n}(\gamma)$ on $\widehat{B}_n \setminus \{(0,0)\}$.

There are some minor redundancies in this list, e.g., the second equation in (9.5) follows from the first. We take $\widehat{\mathcal{R}}^\omega$ to be the space of ψ in $\widehat{\mathcal{R}}^\infty$ that are real-analytic on $\overline{\widehat{B}_n} \setminus \{(0,0)\}$.

The mapping $\rho \mapsto \hat{\rho}$ provides a bijection between $\mathcal{R}^\infty$ and $\widehat{\mathcal{R}}^\infty$, $\mathcal{R}^\omega$ and $\widehat{\mathcal{R}}^\omega$. This follows from (7.2), Proposition 8.1, and easy arguments.

Let $\mathcal{D}^\infty$, $\mathcal{D}^\omega$, $\widehat{\mathcal{D}}^\infty$, $\widehat{\mathcal{D}}^\omega$ denote the spaces of images of B_n, $\widehat{B}_n$ under elements of $\mathcal{R}^\infty$, $\mathcal{R}^\omega$, $\widehat{\mathcal{R}}^\infty$, $\widehat{\mathcal{R}}^\omega$. Thus $\mathcal{D}^\infty$, $\mathcal{D}^\omega$ are spaces of domains in $\mathbf{C}^n$, and $\widehat{\mathcal{D}}^\infty$, $\widehat{\mathcal{D}}^\omega$ are the spaces of $\widehat{D}$ for D in $\mathcal{D}^\infty$, $\mathcal{D}^\omega$.

It is easy and good to represent the $\mathcal{D}$'s as quotients of the $\mathcal{R}$'s. Let $\mathcal{G}^\infty$ denote the set of $\rho \in \mathcal{R}^\infty$ such that $\rho(B_n) = B_n$. From Theorems 3.3 and 3.1 we know that $\mathcal{G}^\infty$ is a group, that $\mathcal{G}^\infty$ acts on $\mathcal{R}^\infty$ by composition on the right, and that two elements of $\mathcal{R}^\infty$ lie in the same coset of $\mathcal{R}^\infty/\mathcal{G}^\infty$ if and only if they have the same image. Thus there is a natural correspondence between $\mathcal{D}^\infty$ and $\mathcal{R}^\infty/\mathcal{G}^\infty$. We define $\mathcal{G}^\omega$, $\widehat{\mathcal{G}}^\infty$, and $\widehat{\mathcal{G}}^\omega$ analogously, and similar remarks hold concerning their actions on $\mathcal{R}^\omega$, $\widehat{\mathcal{R}}^\infty$, $\widehat{\mathcal{R}}^\omega$, and the identification of their quotients with the $\mathcal{D}$'s.

It is helpful, now and in the remaining sections, to keep in mind the $n = 1$ case. When $n = 1$, $\mathcal{D}^\infty$ is the space of smooth simply connected domains in $\mathbf{C}$ that contain the origin, $\mathcal{G}^\infty$ is just the set of rotations of the unit disk, and the identification of $\mathcal{D}^\infty$ with $\mathcal{R}^\infty/\mathcal{G}^\infty$ is classical.

10. SPACES OF RIEMANN MAPPINGS AS COMPLEX VARIETIES

It would be nice if $\mathcal{R}^\infty$ or $\widehat{\mathcal{R}}^\infty$ could be given the structure of a complex Fréchet manifold, or, even better, if the analogue of $\mathcal{R}^\infty$ or $\widehat{\mathcal{R}}^\infty$ based on some other space of functions could be realized as complex Banach or Hilbert manifolds. This is unclear, unfortunately, although we do at least have the (less elegant) results for $\widehat{\mathcal{R}}^\omega$ in Section 8.

It is, however, quite easy to realize the $\mathcal{R}$'s as complex varieties. This indicates that the $\mathcal{R}$'s have complex structures, at least in some sense. We can make this idea reasonably precise by utilizing the complex structure of the ambient vector spaces in which these varieties live. Thus, in order to say that a map defined on $\mathcal{R}^\infty$ is holomorphic, we deem it sufficient that it admit, locally, an extension to the ambient vector space which is holomorphic. We shall discuss this in some detail in the present section, and we shall see that this criterion is adequate to a large degree for our purposes.

We should first be more precise about the realization of the $\mathcal{R}$'s as varieties. In particular we should specify choices for the ambient vector spaces.

Let C_a^∞ denote the vector space of continuous mappings of $\overline{B}_n$ into $\mathbf{C}^n$ that send the origin to itself, are C^∞ on $\overline{B}_n \setminus \{0\}$, and which are holomorphic on $\{\lambda z : \lambda \in \Delta\}$ for all $z \in \partial B_n$. (The "a" stands for "analytic.") This can be viewed as a closed subspace of $C^\infty(\partial B_n, \mathbf{C}^n)$, the space of C^∞ maps of ∂B_n into $\mathbf{C}^n$, in the obvious way. We let C_a^ω denote the subset of C_a^∞ consisting of maps that are real-analytic on $\overline{B}_n \setminus \{0\}$. This is equivalent to requiring real-analyticity on ∂B_n, because of the holomorphicity condition.

Let $\mathcal{O}^\infty$, $\mathcal{O}^\omega$ denote the open subsets of C_a^∞, C_a^ω of mappings that are bilipschitz on $\overline{B}_n$. Thus $\mathcal{R}^\infty$, $\mathcal{R}^\omega$ consist of the elements of $\mathcal{O}^\infty$, $\mathcal{O}^\omega$ which satisfy the homogeneous holomorphic polynomial equation provided by (1.5), i.e., that $\delta\rho(\phi)$ restricts to zero on the subbundle S^2 of the tangent bundle of B_n (defined between (1.3) and (1.4)), where $\phi = dz_1 \wedge \cdots \wedge dz_n$. [In order to truly beat this point to death we should point out that the map from C_a^∞ into complex-valued n-forms on S^2 given by

$$\rho \mapsto \delta\rho(\omega)\,\big|_{S^2}$$

is a homogeneous complex polynomial of degree n, that is, the restriction to the diagonal of a complex n-linear map of $C_a^\infty \times \ldots \times C_a^\infty$ (n times) into the aforementioned space of complex-valued n-forms on S^2. From the analytical point of view it is nice to add that we only need to check that $\delta\rho(\phi)$ vanishes on the restriction of S^2 to ∂B_n; the interior then comes for free, because of the holomorphicity condition on elements of C_a^∞.]

We define $\widehat{C}_a^\infty$, $\widehat{C}_a^\omega$ analogously, but it is a little more complicated this time. As usual we let $\tau_0 : \overline{B}_n \to \overline{\widehat{B}_n}$ be given by $\tau_0(z) = (z, \overline{z})$, and we let Π, Π' denote the projections of $\mathcal{C} = \mathbf{C}^n \times \mathbf{C}^n$ onto its first and second sets of co-ordinates. Let $\widehat{C}_a^\infty$ denote the set of mappings $\psi : \overline{\widehat{B}_n} \to \mathcal{C}$ such that $\Pi \circ \psi \circ \tau_0$ lies in C_a^∞ and $\phi = \Pi' \circ \psi \circ \tau_0$ has the same properties as elements of C_a^∞ do, except that instead of requiring $\phi(\lambda z)$ to be holomorphic in $\lambda \in \Delta$ for all $z \in \partial B_n$, we require that $\lambda\overline{\lambda}^{-1}\phi(\lambda z)$ be holomorphic in λ. Define $\widehat{C}_a^\omega$ similarly, and let $\widehat{\mathcal{O}}^\infty$, $\widehat{\mathcal{O}}^\omega$ be the open subsets of $\widehat{C}_a^\infty$, $\widehat{C}_a^\omega$ of mappings ψ that also satisfy (9.3) and (9.4). The next lemma implies that $\widehat{\mathcal{R}}^\infty$, $\widehat{\mathcal{R}}^\omega$ are subsets of $\widehat{\mathcal{O}}^\infty$, $\widehat{\mathcal{O}}^\omega$.

Lemma 10.1. *If $\psi \in \widehat{\mathcal{R}}^\infty$, then $\Pi \circ \psi \circ \tau_0(\lambda z)$ and $\lambda\overline{\lambda}^{-1}\Pi' \circ \psi \circ \tau_0(\lambda z)$ are holomorphic in $\lambda \in \Delta$ for all $z \in \partial B_n$.*

Thus $\widehat{\mathcal{R}}^\infty$, $\widehat{\mathcal{R}}^\omega$ are the subsets of $\widehat{\mathcal{O}}^\infty$, $\widehat{\mathcal{O}}^\omega$ determined by the constraints in (9.5). Once again these constraints are given in terms of homogeneous holomorphic polynomial mappings on $\widehat{C}_a^\infty$, which are quadratic in this case.

Let us now prove the lemma. This could be accomplished directly from the definitions, but it will be over more quickly if we use our earlier results.

Let $\psi \in \mathcal{R}^\infty$ be given. Proposition 8.1 tells us that $\psi = \hat{\rho}$, which implies that $\rho = \Pi \circ \psi \circ \tau_0$ lies in $\mathcal{R}^\infty \subseteq C_a^\infty$, and that $\Pi' \circ \psi \circ \tau_0$ is given by $\partial F \circ \rho$, where $F = F_0 \circ \rho^{-1}$, as usual.

Consider $\frac{\partial}{\partial\overline{\lambda}}(\partial_j F(\rho(\lambda z)))$ for $\lambda \neq 0$. Because ρ satisfies (1.3), we have

$$\frac{\partial}{\partial\overline{\lambda}}\,(\partial_j F(\rho(\lambda z))) \;=\; \sum_k (\partial_j\overline{\partial}_k F)(\rho(\lambda z))\;\frac{\partial}{\partial\overline{\lambda}}\;\overline{\rho_k(\lambda z)}.$$

Using this and $\delta\rho(\partial\overline{\partial}F) = \partial\overline{\partial}F_0$ it is not difficult to compute that

$$\frac{\partial}{\partial\overline{\lambda}}\,(\partial_j F(\rho(\lambda z))) = \overline{\lambda}^{-1}\partial_j F(\rho(\lambda z)).$$

Hence $\lambda\overline{\lambda}^{-1}(\partial_j F)(\rho(\lambda z))$ is holomorphic for $\lambda \neq 0$. It is holomorphic across $\lambda = 0$ because it is bounded there (and even continuous, with the value zero at the origin). This proves the lemma.

Next we describe the criteria for holomorphicity that we shall use for maps into or from one of the $\mathcal{R}$'s. A map into one of the $\mathcal{R}$'s will be considered holomorphic if it is holomorphic as a map into the corresponding ambient vector space, i.e., C_a^∞ for $\mathcal{R}^\infty$, etc. On the other hand, suppose that we are given a map defined on an open subset of one of the $\mathcal{R}$'s, open relative to the natural topology on the ambient vector space. In order that this map be called holomorphic, we shall consider it sufficient if at each point in the domain of the mapping there is an extension of the mapping to an open set in the ambient vector space which is holomorphic. In the event that the mapping in question is defined on a subset of one of the $\mathcal{R}$'s and takes values in one of the $\mathcal{R}$'s, we will permit the aforementioned extension to take values in the ambient space of the target, and not just the target itself.

Let us consider a simple but important example. Consider the map from $\widehat{\mathcal{R}}^\infty$ (or $\widehat{\mathcal{R}}^\omega$) to $\mathcal{R}^\infty$ ($\mathcal{R}^\omega$) defined by $\psi \mapsto \Pi \circ \psi \circ \tau_0$. The same formula provides an extension of this map from $\widehat{C}_a^\infty$ to C_a^∞ (or $\widehat{C}_a^\omega$ into C_a^ω) which is holomorphic, and in fact complex linear. According to the conventions above, we consider this to be a holomorphic map from the $\widehat{\mathcal{R}}$'s to the $\mathcal{R}$'s.

What about its inverse mapping, $\rho \mapsto \hat{\rho}$? This turns out to be more complicated, but we do have the following.

Proposition 10.2. *Fix $\rho' \in \mathcal{R}^\infty$. Then for $\epsilon > 0$ small enough there is a holomorphic map of $\{\rho \in C_a^\infty : \|d\rho - d\rho'\|_{L^\infty(\partial B_n)} < \epsilon\}$ into $\widehat{C}_a^\infty$ that restricts to $\rho \mapsto \hat{\rho}$ on $\mathcal{R}^\infty$.*

Thus, according to our conventions, $\rho \mapsto \hat{\rho}$ determines a holomorphic map of $\mathcal{R}^\infty$ into $\widehat{\mathcal{R}}^\infty$. The proof of this proposition doesn't work in the real-analytic case, because it uses partitions of unity, but it would not be surprising to learn that a similar result still holds.

There is a technical point that arises in the proof of the proposition of somewhat broader significance, and so we discuss it first. In proving Proposition 10.2 it is easier to produce a holomorphic map that takes values in $C^\infty(\partial\widehat{B}_n, \mathcal{C})$ rather than $\widehat{C}_a^\infty$. It would not be unreasonable to enlarge our conventions for holomorphicity so that a map from $\mathcal{R}^\infty$ to $\widehat{\mathcal{R}}^\infty$ is considered to be holomorphic if it admits a holomorphic extension to a map from C_a^∞ to $C^\infty(\partial\widehat{B}_n, \mathcal{C})$, or to make similar expansions of the previously stated criteria for holomorphicity in other cases. In the present situation this is not necessary because of the next lemma, which will be helpful for dealing with this issue in general.

Lemma 10.3. *There are continuous complex-linear projections of $C^\infty(\partial B_n, \mathbf{C}^n)$ onto C_a^∞, $C^\omega(\partial B_n, \mathbf{C}^n)$ onto C_a^ω, $C^\infty(\partial\widehat{B}_n, \mathcal{C})$ onto $\widehat{C}_a^\infty$, and $C^\omega(\partial\widehat{B}_n, \mathcal{C})$ onto $\widehat{C}_a^\omega$.*

The proof is effected through a fairly straightforward use of Cauchy integrals. Given $f \in C^\infty(\partial B_n, \mathbf{C}^n)$, define $P(f)$ on B_n by

$$P(f)(\lambda z) = \frac{1}{2\pi i} \int_{|\alpha|=1} f(\alpha z) \left\{ \frac{1}{\alpha - \lambda} - \frac{1}{\alpha} \right\} d\alpha \tag{10.4}$$

for all $\lambda \in \Delta$, $z \in \partial B_n$. It is not difficult to show that this is well defined, i.e., that if $\lambda z = \lambda' z'$, then you get the same value for $P(f)$. We take

$$P(f)(z) = \lim_{\lambda \to 1} P(f)(\lambda z) \qquad \text{for } z \in \partial B_n.$$

Well-known results imply that this limit always exists. It is not hard to verify that P provides a continuous complex-linear projection from $C^\infty(\partial B_n, \mathbf{C}^n)$ to C_a^∞, and also from $C^\omega(\partial B_n, \mathbf{C}^n)$ to C_a^ω.

The story with $\hat{}$'s is similar. Let f be an element of $C^\infty(\partial\widehat{B}_n, \mathcal{C})$, and let $f_1 = \Pi \circ f \circ \tau_0$, $f_2 = \Pi' \circ f \circ \tau_0$, where $\tau_0 : B_n \to \widehat{B}_n$ is given by $\tau_0(z) = (z, \overline{z})$ as usual. We define our projection Q by $(Qf) \circ \tau_0 = (P(f_1), P'(f_2))$, where $P'(f_2)$ is given by

$$P'(f_2)(\lambda z) = \frac{\overline{\lambda}}{\lambda} \frac{1}{2\pi i} \int_{|\alpha|=1} f_2(\alpha z) \frac{\alpha}{\overline{\alpha}} \left\{ \frac{1}{\alpha - \lambda} - \frac{1}{\alpha} \right\} d\alpha \tag{10.5}$$

for $\lambda \in \Delta$, $z \in \partial B_n$, and by

$$P'(f_2)(z) = \lim_{\lambda \to 1} P'(f_2)(z)$$

for $z \in \partial B_n$. Once again it is easy to check that (10.5) is well-defined in the sense that $P'(f_2)(\lambda z) = P'(f_2)(\tilde{\lambda}\tilde{z})$ as long as $\lambda z = \tilde{\lambda}\tilde{z}$. It is also not hard to check that Q defines a continuous complex-linear projection from $C^\infty(\partial\widehat{B}_n, \mathcal{C})$ to $\widehat{C}_a^\infty$, and from $C^\omega(\partial B_n, \mathcal{C})$ to $\widehat{C}_a^\omega$, as desired.

The proof of Proposition 10.2 is a little tricky. Recall that $\hat{\rho} : \overline{\widehat{B}}_n \to \mathcal{C}$ is defined in terms of ρ by

$$\hat{\rho} = (\rho \circ \Pi,\ \partial F \circ \rho \circ \Pi),$$

where $F = F_0 \circ \rho^{-1}$, as always. We need to find a formula for $(\partial F) \circ \rho$ in terms of ρ that admits a holomorphic extension to an open set in C_a^∞.

From Theorem 2.2 we know that $\delta\rho(\partial F) = \partial F_0$ if $\rho \in \mathcal{R}^\infty$, so that

$$\partial F_0 = \sum_{jk} ((\partial_k F) \circ \rho) \left\{ (\partial_j \rho_k) dz_j + (\overline{\partial}_j \rho_k) d\overline{z}_j \right\},$$

where ρ_k, $k = 1, \ldots, n$, denote the components of ρ with respect to the standard basis on $\mathbf{C}^n$. Because the left side is a $(1,0)$ form, the $d\overline{z}_j$ terms must disappear, leaving

$$\partial_j F_0 = \sum_k (\partial_k F) \circ \rho\ (\partial_j \rho_k). \tag{10.6}$$

We would like to invert $\partial\rho$ to get a nice formula for ∂F, but we have no reason to know that $\partial\rho$ will always be invertible for $\rho \in \mathcal{R}^\infty$. Instead we shall show that for ρ near ρ' we can replace $\partial\rho$ by something which is invertible and has other desirable features.

Fix $\rho' \in \mathcal{R}^\infty$. Let $\{b_l\}$ be a finite covering of ∂B_n by open balls of radius $\delta > 0$, where δ is small enough so that the following is true. For each l there are $n-1$ $\mathbf{C}^n$-valued smooth functions $w_{l,j}(z)$, $1 \le j \le n-1$, defined on $\partial B_n \cap \overline{b}_l$, such that:

(10.7) $w_{l,j}(z) \in S_z^2$ for all $z \in \partial B_n \cap \overline{b}_l$, where S_z^2 is as in Section 1;

(10.8) for each l and each z, $w_{l,j}(z)$, $1 \le j \le n-1$, are linearly independent over $\mathbf{C}$;

(10.9) for each l and each z, $d\rho'_z(w_{l,j}(z))$, $1 \le j \le n-1$, are also linearly independent over $\mathbf{C}$.

It is easy to see that such functions $w_{l,j}$ exist if δ is small enough. Notice that the complex linear span of $\{w_{l,j}(z)\}_j$ and $\{d\rho'_z(w_{l,j}(z))\}_j$ are S_z^2 and $d\rho'_z(S_z^2)$, respectively. For the latter we use the fact that $\rho' \in \mathcal{R}^\infty$ and so satisfies (1.4).

Given $\rho \in C_a^\infty$, l, and $z \in \partial B_n \cap \overline{b}_l$, let $M_l(\rho)_z$ denote the complex-linear transformation on $\mathbf{C}^n$ that agrees with $d\rho$ on S_z^1 and on $w_{l,j}(z)$, $1 \le j \le n-1$, and let $(M_l(\rho)_z)_{jk}$ denote the matrix of $M_l(\rho)_z$ with respect to the standard basis on $\mathbf{C}^n$. It is easy to see that $M_l(\rho)_z$ depends complex-linearly on ρ. If $\|d\rho - d\rho'\|_{L^\infty(\partial B_n)} \le \epsilon$

for $\epsilon > 0$ small enough, then $d\rho_z(w_{l,j}(z))$, $1 \leq j \leq n-1$, are linearly independent over $\mathbf{C}$ for all l and $z \in \partial B_n \cap \overline{b}_l$, and hence $M_l(\rho)_z$ is invertible. We assume from now on that ϵ has been chosen so that this is true.

If $\rho \in \mathcal{R}^\infty$ and $\|d\rho - d\rho'\|_{L^\infty(\partial B_n)} \leq \epsilon$, then $d\rho_z$ and $M_l(\rho)_z$ both map S_z^2 to the same subspace of $\mathbf{C}^n$, for all l and all $z \in \partial B_n \cap \overline{b}_l$. [Indeed, the image of S_z^2 under $d\rho_z$ is a complex subspace, since ρ satisfies (1.4), and so it must contain $M_l(\rho)_z(S_z^2)$, by definitions. Both $d\rho_z(S_z^2)$ and $M_l(\rho)_z(S_z^2)$ have complex dimension $n-1$, and so they must be equal.] We also have that $\partial F \big|_{\rho(z)}$ vanishes on $d\rho_z(S_z^2)$, since $\delta\rho(\partial F) = \partial F_0$ and $\partial F_0 \big|_z$ vanishes on S_z^2. Therefore (10.6) implies that

$$\partial_j F_0(z) = \sum_k \partial_k F(\rho(z))(M_l(\rho)_z)_{jk} \quad \text{for } z \in \overline{b}_l. \tag{10.10}$$

Let us use this to extend $\rho \mapsto \hat{\rho}$ to $\rho \in C_a^\infty$ near ρ'. Let ϕ_l be a C^∞ partition of unity on ∂B_n subordinate to the covering by the b_l's. For $\rho \in C_a^\infty$ such that $\|d\rho - d\rho'\|_{L^\infty(\partial B_n)} \leq \epsilon$, consider the mapping of ∂B_n into $\mathcal{C}$ defined by

$$\left(\rho, \left\{\sum_l \phi_l \sum_j (\partial_j F_0)\,(M_l(\rho))^{jk}\right\}_{k=1}^n\right)$$

where $(M_l(\rho))^{jk}$ denotes the matrix of the inverse of $M_l(\rho)$. [We suppress the dependence on z in this notation.] This formula provides a holomorphic map of $\{\rho \in C_a^\infty : \|d\rho - d\rho'\|_{L^\infty(\partial B_n)} \leq \epsilon\}$ into $C^\infty(\partial B_n, \mathcal{C})$ that agrees with $\rho \mapsto \hat{\rho} \circ \tau_0$ when $\rho \in \mathcal{R}^\infty$, because of (10.10). This is not quite what we wanted, since the range of this map is not $\widehat{C}_a^\infty$, but that is easily rectified using Lemma 10.3.

11. HOMOGENEOUS MAPPINGS, COMPLETELY CIRCLED DOMAINS, AND THE KOBAYASHI INDICATRIX

In this section we are going to look at the topics just listed in the context of spaces of Riemann mappings and spaces of domains.

Let $\mathcal{S}^\infty$, $\mathcal{S}^\omega$ denote the subsets of $\mathcal{R}^\infty$, $\mathcal{R}^\omega$ of mappings ρ that are complex homogeneous of degree 1. Let C_h^∞, C_h^ω denote the set of continuous maps from $\overline{B}_n$ to $\mathbf{C}^n$ that are complex homogeneous of degree 1 and C^∞, C^ω away from the origin. Thus $C_h^\infty \subseteq C_a^\infty$, $C_h^\omega \subseteq C_a^\omega$ a fortiori.

Define $S : \mathcal{R}^\infty \to \mathcal{S}^\infty$ by $S(\rho) = \sigma$, where σ is given by (4.3), or, equivalently, (4.4) or (4.6). It follows from Theorem 4.2 and (4.6) that S does indeed map $\mathcal{R}^\infty$ into $\mathcal{S}^\infty$, and that S sends $\mathcal{R}^\omega$ to $\mathcal{S}^\omega$. Because of (4.6) it is apparent that S extends to a continuous complex-linear mapping of C_a^∞ onto C_h^∞, as well as C_a^ω onto C_h^ω, and therefore S is holomorphic, in accordance with the conventions of Section 10.

Similarly, we define $\widehat{\mathcal{S}}^\infty$, $\widehat{\mathcal{S}}^\omega$ to be the set of ψ in $\widehat{\mathcal{R}}^\infty$, $\widehat{\mathcal{R}}^\omega$ that commute with δ_λ for all $\lambda \in \Delta$, where $\delta_\lambda : \mathcal{C} \to \mathcal{C}$ is defined for $\lambda \in \mathbf{C}$ by $\delta_\lambda((z,\zeta)) = (\lambda z, \overline{\lambda}\zeta)$. There is the same kind of correspondence between the $\mathcal{S}$'s and the $\widehat{\mathcal{S}}$'s as there is for the $\mathcal{R}$'s; if σ lies in $\mathcal{S}^\infty$ (or $\mathcal{S}^\omega$), then $\hat{\sigma}$ (as defined by (7.1)) lies in $\widehat{\mathcal{S}}^\infty$ ($\widehat{\mathcal{S}}^\omega$), and conversely, $\psi \mapsto \Pi \circ \psi \circ \tau_0$ sends $\widehat{\mathcal{S}}^\infty$, $\widehat{\mathcal{S}}^\omega$ to $\mathcal{S}^\infty$, $\mathcal{S}^\omega$. Let $\widehat{C}_h^\infty$, $\widehat{C}_h^\omega$ denote the spaces of continuous mappings from $\overline{\widehat{B}}_n$ to $\mathcal{C}$ that are C^∞, C^ω away from $(0,0)$ and which commute with δ_λ for all $\lambda \in \Delta$. We define $\widehat{S} : \widehat{\mathcal{R}}^\infty \to \widehat{\mathcal{S}}^\infty$ by

$$\widehat{S}(\psi) = \lim_{t \to 0} t^{-1}(\psi \circ \delta_t), \quad t > 0, \tag{11.1}$$

and it is not hard to see that $\widehat{S}$ sends $\widehat{\mathcal{R}}^\infty$ into $\widehat{\mathcal{S}}^\infty$ and $\widehat{\mathcal{R}}^\omega$ into $\widehat{\mathcal{S}}^\omega$. This uses a formula for $\widehat{S}(\psi)$ based on Cauchy integrals that is analogous to (4.6). [If you want to write this formula in detail, it is helpful to consult the reproducing formula for $\widehat{C}_a^\infty$ given in the proof of Lemma 10.3.] This analogue of (4.6) also provides a continuous complex-linear extension of $\widehat{S}$ to a map from $\widehat{C}_a^\infty$ into $\widehat{C}_h^\infty$, and this extension sends $\widehat{C}_a^\omega$ into $\widehat{C}_h^\omega$ as well. Thus $\widehat{S}$ is holomorphic according to Section 10.

Let K denote the operation that assigns to a domain in $\mathbf{C}^n$ its Kobayashi indicatarix, defined by (4.1), and define $R : \mathcal{R}^\infty \to \mathcal{D}^\infty$ by $R(\rho) = \rho(B_n)$. ["R" stands for "range."]

Theorem 11.2. *As operations on* $\mathcal{R}^\infty$,

$$K \circ R = R \circ S. \tag{11.3}$$

In other words, if $D = \rho(B_n)$, $\rho \in \mathcal{R}^\infty$, and $\sigma = S(\rho)$, then $\sigma(B_n)$ is the Kobayashi indicatrix of D. This is part of Theorem 4.2.

Let $\mathcal{I}^\infty$ and $\mathcal{I}^\omega$ denote the sets of completely circled, strongly pseudoconvex domains in $\mathbf{C}^n$ with C^∞, C^ω boundary. Then $\mathcal{I}^\infty$, $\mathcal{I}^\omega$ are subsets of $\mathcal{D}^\infty$, $\mathcal{D}^\omega$, because of Theorem 3.4 and its real-analytic version. They are also the images of $\mathcal{D}^\infty$, $\mathcal{D}^\omega$ under K.

Now let's look at the space of domains with prescribed indicatrix. For simplicity of exposition we restrict ourselves to the C^∞ category.

Fix $I \in \mathcal{I}^\infty$, and set $\mathcal{D}^\infty(I) = K^{-1}(I) \cap \mathcal{D}^\infty$. Let σ be any element of $\mathcal{S}^\infty$ with $\sigma(B_n) = I$, and set $\mathcal{R}^\infty(\sigma) = S^{-1}(\sigma) \cap \mathcal{R}^\infty$. It is not hard to check that R restricts to a bijection between $\mathcal{R}^\infty(\sigma)$ and $\mathcal{D}^\infty(I)$. [Injectivity follows from Theorems 3.1(a) and 3.3. As for surjectivity, suppose that $D \in \mathcal{D}^\infty(I)$, so that $D = \rho(B_n)$ for some $\rho \in \mathcal{R}^\infty$. If $S(\rho) \neq \sigma$, then you can compose ρ on the right by $S(\rho)^{-1} \circ \sigma \in \mathcal{G}^\infty$ to obtain D as the range of an element of $\mathcal{R}^\infty(\sigma)$.]

This allows us to identify $\mathcal{D}^\infty(I)$ with a complex variety in $\mathcal{O}^\infty$, namely the intersection of $\mathcal{R}^\infty$ with the complex affine subspace $\{\rho \in C_a^\infty : S(\rho) = \sigma\}$. Different choices of σ produce different representations of $\mathcal{D}^\infty(I)$ as a variety, but they are all equivalent, in the following sense. If $\tilde{\sigma}$ is another element of $\mathcal{S}^\infty$ with $\tilde{\sigma}(B_n) = I$, then $\tilde{\sigma} = \sigma \circ \alpha$ for some $\alpha \in \mathcal{G}^\infty$, and the map $\rho \mapsto \rho \circ \alpha$ is a complex-linear isomorphism of C_a^∞ onto itself that takes $\mathcal{R}^\infty(\sigma)$ to $\mathcal{R}^\infty(\tilde{\sigma})$.

Similarly, we could identify $\mathcal{D}^\infty(I)$ with $\widehat{\mathcal{R}}^\infty \cap \widehat{S}^{-1}(\hat{\sigma})$, which is a complex variety in $\widehat{\mathcal{O}}^\infty$, and again there is the same sort of uniqueness statement. We could also identify $\mathcal{D}^\omega(I)$ with $\widehat{\mathcal{R}}^\omega \cap \widehat{S}^{-1}(\hat{\sigma})$ when σ and I are real analytic. This has the advantage of being amenable to analysis by the method of generating functions as discussed in Section 8. In particular the action of $\widehat{S}$ on $\widehat{\mathcal{R}}^\omega$ cooperates well with the generating functions story.

Although $\mathcal{D}^\infty(I)$ admits a natural representation as a complex variety for each $I \in \mathcal{I}^\infty$, the same is not true of $\mathcal{D}^\infty$ itself. It is particularly clear that $\mathcal{D}^\infty$ has no natural complex structure when $n = 1$; in that case the analogue of $\mathcal{D}^\infty(I)$ has real codimension 1 inside $\mathcal{D}^\infty$.

In this regard it is good to think of $\mathcal{D}^\infty$ as being $\mathcal{R}^\infty/\mathcal{G}^\infty$. Although $\mathcal{R}^\infty$ is a complex variety, $\mathcal{G}^\infty$ is not a complex group, and this is what causes us to lose the "complex structure" when we pass to the quotient. We can think of the mapping $R : \mathcal{R}^\infty \to \mathcal{D}^\infty$ as resolving, in some sense, the noncomplex directions of $\mathcal{D}^\infty$, and we can try to use this to endow $\mathcal{D}^\infty$ with some geometric structure that substitutes for the lack of a complex structure.

One way to think of a geometric structure is as something on a space that distinguishes a family of mappings into that space. For Riemannian structures, for instance, we have geodesics and harmonic mappings, while for complex structures we have holomorphic mappings. There is an obvious class of distinguished mappings in $\mathcal{D}^\infty$, to wit, those mappings T that can be factored as $R \circ \widetilde{T}$, where $\widetilde{T}$ takes values in $\mathcal{R}^\infty$ and is holomorphic as a map into C_a^∞. [We assume here that the domain of T is such that it makes sense to talk about the holomorphicity of $\widetilde{T}$.] Let us call such a mapping "special". More generally it is convenient to call T special if at each point in its domain there is a neighborhood on which such a factorization as above exists, so that the property of being special is a local one.

Special mappings into $\mathcal{D}^\infty$ have some nice properties.

Lemma 11.4. *If T is a special mapping into $\mathcal{D}^\infty$, then $K \circ T$ is also special.*

There are two obvious interpretations of the statement that $K \circ T$ is special, and they are equivalent. The first is that $K \circ T$ is special as a mapping into $\mathcal{D}^\infty$ that just so happens to take values in $\mathcal{I}^\infty$. The second is that $K \circ T$ admits, locally, a factorization into a holomorphic map into $\mathcal{S}^\infty$ followed by R. The equivalence is easily verified, by chasing definitions. The lemma follows immediately from Theorem 11.2.

Theorem 11.5. *Suppose that $\alpha \mapsto I(\alpha)$ is a smooth map from some domain U in $\mathbf{C}^m$ into $\mathcal{I}^\infty$. Let $F_{I(\alpha)}$ be associated to each $I(\alpha)$ as in Section 3, and define $G : U \times \mathbf{C}^n \to \mathbf{R}$ by $G(\alpha, z) = F_{I(\alpha)}(z)$. Then $\alpha \mapsto I(\alpha)$ is special if and only if $(\partial\overline{\partial}G)^{n+1} \equiv 0$ on $U \times (\mathbf{C}^n \setminus \{0\})$.*

This result has been stated in the C^∞ category only for notational convenience, and it is easy to generalize the theorem and its proof to the case where you have only a few derivatives (like 3). However, a loss of smoothness does occur, and so the results in that case are more awkward to state.

If the $I(\alpha)$'s are convex domains then there are some alternative characterizations of special mappings that are quite interesting, particularly for the purpose of having interpretations of special mappings in terms of complex analysis. For this discussion let us take $m = 1$ and $U =$ the unit disk, and assume that $\alpha \mapsto I(\alpha)$ extends continuously to $\overline{U}$. If the $I(\alpha)$'s are convex, then $\alpha \mapsto I(\alpha)$ is special if and only if the family of Banach space norms on $\mathbf{C}^n$ with the $I(\alpha)$'s as unit balls is a complex interpolation family in the sense of [C²RSW]. This follows from Theorem 11.5 and a result of Rochberg [R]. Another equivalent condition is the requirement that

$$\{(\alpha, z) \in \mathbf{C} \times \mathbf{C}^n : |\alpha| \leq 1, z \in I(\alpha)\}$$

be the polynomial hull of

$$\{(\alpha, z) \in \mathbf{C} \times \mathbf{C}^n : |\alpha| = 1, z \in I(\alpha)\}.$$

(More information about the relationship between polynomial hulls and interpolation of Banach spaces can be found in [Sl].)

We shall see that Theorem 11.5 is really just a variation of Theorem 8.11 in [S]. To do this we must first reconcile the notations and such employed here with those in [S], especially Sections 7 and 8. This will be helpful in later sections also, where other connections with [S] will arise.

In [S] we were working on a general complex manifold X rather than simply $\mathbf{C}^n$, and so for our purposes we simply set $X = \mathbf{C}^n$. Unfortunately most of [S] was written in such a way as to accomodate nontrivial topology on X, which adds complications that are irrelevant in the present situation. For this reason some things will be handled differently here than in [S]. Some of these differences are discussed in Section 9 in [S].

It should be be kept in mind that although we are taking $X = \mathbf{C}^n$, we are for the moment restricting ourselves to objects that are homogeneous, which was generally not the case in [S]. Again, this necessitates some modifications in the story, some of which are discussed in Section 9 in [S].

Fortunately, $\mathcal{C}$ was used to denote the complex cotangent bundle of X in [S], but it is better to make some changes in the definitions of M and $\widehat{M}$ from [S], owing to the lack of nontrivial topology of $X = \mathbf{C}^n$. We take $\widehat{M}$ to be $\mathcal{C}$ both as a set and with the same complex structure, and we take $M = \{(z, \overline{z}) \in \widehat{M} : z \in \mathbf{C}^n\}$. In [S] we took M to be the zero section of $\mathcal{C}$, but we gave $\widehat{M}$ a different complex structure from that of $\mathcal{C}$. We identify X as a smooth manifold with M, but of course M is totally real in $\widehat{M}$, and so this identification is not compatible with the complex structure on X.

Define $\widehat{\Sigma}_h$ to be the set of continuous mappings $\psi : M \to \widehat{M}$ which are C^∞ embeddings away from $(0,0)$, which commute with δ_λ for all $\lambda \in \mathbf{C}$, and which satisfy $\delta\psi(\gamma) = \delta i_M(\gamma)$ away from $(0,0)$. This is pretty much the same thing as what was called $\widehat{\Sigma}$ in [S], except for the homogeneity condition $\psi \circ \delta_\lambda = \delta_\lambda \circ \psi$, which is the reason for the subscript "h". Let $\mathcal{G}\widehat{\Sigma}_h$ ("$\mathcal{G}$" for "graph") denote the set of ψ's in $\widehat{\Sigma}_h$ such that $\Pi \circ \psi$ is bilipschitz.

Lemma 11.6. *$\psi \in \mathcal{G}\widehat{\Sigma}_h$ if and only if $\psi \mid_{\widehat{B}_n} \in \widehat{\mathcal{S}}^\infty$.*

Of course $\psi \in \widehat{\Sigma}_h$ is determined by $\psi \mid_{\widehat{B}_n}$, because of homogeneity.

Before proving this lemma, and then Theorem 11.5, we make a couple of remarks.

There is nothing special about the C^∞ category in this lemma; the proof will work as long as ψ is C^1 away from $(0,0)$, say.

Consider for example $\widehat{\mathcal{S}}(H^s)$, which is defined in the same way that $\widehat{\mathcal{S}}^\infty$ was, except that we replace the condition that ψ be C^∞ away from $(0,0)$ with the requirement that ψ belong locally to the Sobolev space H^s away from $(0,0)$. If $s > n$, then $\widehat{\mathcal{S}}(H^s)$ is a complex Hilbert manifold. Using Lemma 11.6 this reduces to a minor variation of Proposition 7.2 in [S], which is itself a minor variation of a result in [EM]. You have to make some changes in the proof of this proposition to keep track of the homogeneity, and you must also use the homogeneity to make up for the lack of compactness of the domain manifold.

This complex Hilbert manifold structure in $\widehat{\mathcal{S}}(H^s)$ is very helpful. It tells us in particular that there are lots of holomorphic maps into $\widehat{\mathcal{S}}(H^s)$, for s as large as we like, and hence that there are plenty of special maps into $\mathcal{I}(H^s)$ by Theorem 11.5.

Let us now prove Lemma 11.6. The "if" part is immediate from the definitions. Let us check the "only if" part.

Suppose that $\psi \in \widehat{\Sigma}_h$ and that $\Pi \circ \psi$ is bilipschitz. As usual define $\tau_0 : \mathbf{C}^n \to \mathcal{C}$ by $\tau_0(z) = (z, \overline{z})$. Set $\rho = \Pi \circ \psi \circ \tau_0$, so that ρ is complex homogeneous of degree 1.

Let $\beta : \mathbf{C}^n \to \mathbf{C}^n$ be the function determined by $\psi(\tau_0(z)) = (\rho(z), \beta(\rho(z)))$. There is a real-valued function F on $\mathbf{C}^n$ such that $\partial_j F = \beta_j$ away from the origin, since the image of ψ is ν-Lagrangian and a graph. [This uses $\delta\psi(\gamma) = \delta i_M(\gamma)$ and the fact that M is ν-Lagrangian, as well as (6.4) and (6.6). Notice that we can take F to be defined and locally Lipschitz on $\mathbf{C}^n$, and not just $\mathbf{C}^n \backslash \{0\}$; we first produce it on $\mathbf{C}^n \setminus \{0\}$, by the appropriate integration, and then use $\partial_j F = \beta_j \in L^\infty_{\text{loc}}$ to conclude that F is locally Lipschitz on $\mathbf{C}^n \setminus \{0\}$ and has a Lipschitz extension across the origin.] If we can show that $\rho\,|_{B_n}$ lies in $\mathcal{S}^\infty$, and that $F = F_0 \circ \rho^{-1}$, then it follows that $\psi\,|_{\widehat{B}_n}$ lies in $\widehat{\mathcal{S}}^\infty$.

If we choose F so that it vanishes at the origin, then

$$F(\lambda z) = |\lambda|^2 F(z) \tag{11.7}$$

for all $\lambda \in \mathbf{C}$, $z \in \mathbf{C}^n$. This follows from $\beta(\lambda z) = \overline{\lambda}\beta(z)$, which is itself a consequence of the assumption that ψ commutes with δ_λ. We assume from now on that F has this property.

Define $\tau : \mathbf{C}^n \to \mathcal{C}$ by $\tau(t) = (z, \beta(z))$. From (6.5) and (6.7) we have

$$\delta\tau(\Gamma) = \partial F,\ \delta\tau(\mu) = \frac{1}{2i}\,\partial\overline{\partial}F$$

away from the origin, and similarly $\delta\tau_0(\Gamma) = \partial F_0$, $\delta\tau_0(\mu) = \frac{1}{2i}\partial\overline{\partial}F_0$. Since $\delta\psi(\gamma) = \delta i_M(\gamma)$ we obtain

$$\delta\rho\left(\frac{1}{2i}\,\partial\overline{\partial}F\right) = \delta\rho\,(\delta\tau(\mu)) = \delta(\psi \circ \tau_0)(\mu) = \delta\tau_0(\mu) = \frac{1}{2i}\,\partial\overline{\partial}F_0 \tag{11.8}$$

away from 0.

Next we observe that

$$\delta\rho(\partial F) = \partial F_0. \tag{11.9}$$

This is proved in the same way as the "if" part of Theorem 3.3. [Namely, you use (11.7) to express ∂F, ∂F_0 in terms of contractions of $\partial\overline{\partial}F$, $\partial\overline{\partial}F_0$ by certain vector fields that are preserved by ρ, because it is complex homogeneous of degree 1, and then you use (11.8).] This implies that $\rho\,|_{B_n} \in \mathcal{S}^\infty$ by Theorem 2.2. Because $\delta\rho(dF) = dF_0$ (by (11.9)) and $F(0) = 0 = F_0(0)$, we have that $F = F_0 \circ \rho^{-1}$, and this implies that $\psi\,|_{\widehat{B}_n}$ lies in $\widehat{\mathcal{S}}^\infty$, as we noted earlier. This finishes the proof of Lemma 11.6.

Let us now derive Theorem 11.5 from Theorem 8.11 in [S]. There is still a little more notation that has to be explained. The two-form ω on X in [S] should be taken to be $\frac{1}{2i}\partial\overline{\partial}F_0$ in our case. The variable z in [S] plays the role that α does here, although in [S] we restricted ourselves to $m = 1$. The family of 1-forms $\phi(\alpha, \cdot)$ on X should be taken to be $\partial F_{I(\alpha)} - \partial F_0$. [As mentioned previously, we had to work with 1-forms in [S] instead of merely functions to accomodate nontrivial topology.] However, $\mathcal{L}(\phi(\alpha,\cdot))$ should not be taken to be the graph of $\phi(\alpha,\cdot)$ in $\mathcal{C}$, as in [S], but rather the graph of $\partial F_{I(\alpha)}$. This is another manifestation of the fact that we are using the complex structure of $\mathcal{C}$ on $\widehat{M}$, and that we are letting M be $\{(z,\overline{z}) \in \widehat{M} : z \in \mathbf{C}^n\}$, rather than using a different complex structure on $\widehat{M}$, and taking M to be the zero section of $\mathcal{C}$, as in [S].

Consider first the "only if" part of Theorem 11.5. It is easy to reduce to the case where $m = 1$. [For this it is helpful to remember that $F_{I(\alpha)}(z)$ must be strictly plurisubharmonic in z away from $z = 0$ for each α.]

Suppose that $\alpha \mapsto I(\alpha)$ is special. This means that we can find, about a neighborhood of any given α_0, a mapping $\alpha \mapsto \sigma_\alpha$ into $\mathcal{S}^\infty$ that is holomorphic in α and satisfies $\sigma_\alpha(B_n) = I(\alpha)$. Homogeneity then forces $F_{I(\alpha)} = F_0 \circ \sigma_\alpha^{-1}$. If we define $\hat{\sigma}_\alpha \in \widehat{\mathcal{S}}^\infty$ by (7.1), as usual, then $\hat{\sigma}_\alpha$ also depends holomorphically on α, because of Proposition 10.2.

Let $f(\alpha, \cdot)$ be the map of M into $\widehat{M}$ that commutes with δ_λ and agrees with $\hat{\sigma}_\alpha$ on $\widehat{B}_n$. From Lemma 11.6 it follows that $f(\alpha, \cdot) \in \mathcal{G}\widehat{\Sigma}_h$ for all α, and of course it depends holomorphically on α.

At this point we want to invoke the "if" part of Theorem 8.11 in [S]. Although X was assumed to be compact in that theorem, this requirement played no role in the proof of the "if" part. The conclusion of that theorem is that $\phi(\alpha, \cdot)$ satisfies a certain equation which is almost equivalent to $(\partial\overline{\partial}G)^{n+1} \equiv 0$. In our case it is equivalent; the homogeneity conditions on $F_{I(\alpha)}$ prevent the small ambiguities in choosing potentials for the $\phi(\alpha, \cdot)$'s that could otherwise occur.

It remains to verify the "if" part of Theorem 11.5. Assume at first that $m = 1$. We want to apply the "only if" part of Theorem 8.11 in [S]. If we can do that then we get that for each simply-connected region U_0 in $U \subseteq \mathbf{C}$ we can find a map $f(\alpha, w)$ of $U_0 \times \mathbf{C}^n$ into $\mathcal{C}$ which is holomorphic in α and whose image is the graph of $\partial F_{I(\alpha)}$.

The problem with applying Theorem 8.11 in [S] is that compactness of X was assumed, and needed, for this part. It was needed to pass from local existence of $f(\alpha, w)$ to global existence, in the following two ways: first, if you fix α_0, then there is a neighborhood of α_0 such that $f(\alpha, w)$ is defined for all α in that neighborhood and all w; and second, that $f(\alpha, w)$ is in fact globally defined on any simply connected region of α's. For the first issue homogeneity easily substitutes for compactness. The second issue doesn't really matter for proving Theorem 11.5, because of the way it was stated, but let's address it anyway.

The main point was to know that $g(\alpha, w)$ in (8.15) in [S] could not blow up. In our case the homogeneity can be used to show that $F_{I(\alpha)}(g(\alpha, w))$ is constant in w, which does the job.

[Let us sketch the proof of this conservation rule. For each $w \in \mathbf{C}^n$, $w \neq 0$, $(\alpha, g(\alpha, w))$ lies on a single leaf of the foliation of $U \times (\mathbf{C}^n \setminus \{0\})$ determined by the (real) rank 2 subbundle of the tangent bundle of $U \times (\mathbf{C}^n \setminus \{0\})$ defined by

$$\{v \in T_{(\alpha,w)}(U \times (\mathbf{C}^n \setminus \{0\})) : \, i(v)\theta \,|_{(\alpha,w)} = 0\},$$

where θ denotes the two-form $\frac{1}{2i}\partial\overline{\partial}G$ on $U \times (\mathbf{C}^n \setminus \{0\})$. This is how $g(\alpha, w)$ was built in [S]. It is not hard to show that $\frac{\partial}{\partial \alpha}(F_{I(\alpha)}(g(\alpha, w))) = 0$ using this, the fact that $g(\alpha, w)$ is holomorphic in α, and the homogeneity identity

$$\sum_l \left(\partial_j \overline{\partial}_l F_{I(\alpha)}\right)(z)\overline{z}_l = \partial_j F_{I(\alpha)}(z)$$

(as in (3.9)).]

Thus we can indeed apply Theorem 8.11 in [S], even though X is not compact. The construction in the proof also gives that $f(\alpha, w)$ satisfies the homogeneity condition

$$f(\alpha, \lambda w) = \delta_\lambda(f(\alpha, w))$$

for all $\lambda \in \mathbf{C}$, and that $f(\alpha, \cdot)$ is a smooth embedding of $\mathbf{C}^n \setminus \{0\}$ into $\mathcal{C}$ for each α. It is also bilipschitz across $w = 0$.

The conclusion of the "only if" part of Theorem 8.11 in [S] is not quite as strong as we need here, but it is easily strengthened, as observed in [S]. Set $f_\alpha = f(\alpha, \cdot)$. Then $\delta f_\alpha(\gamma) = \delta f_\alpha(\mu)$, because the image of f_α is ν-Lagrangian. The fact that f_α is holomorphic in α, and that γ is a holomorphic (2,0)-form on $\mathcal{C}$, forces $\delta f_\alpha(\mu)$ to be constant in α. [This can also be verified directly from the construction of f_α.] As pointed out in [S], we can assume that $\delta f_\alpha(\mu) = \frac{1}{2i}\partial\overline{\partial}(F_{I(\alpha_0)})$ for some fixed α_0; we can always reduce to this case by composing the f_α's by a fixed mapping on the right (namely, the inverse of $\Pi \circ f_{\alpha_0}$ — see (6.7)). By Theorem 3.4 there is a map ρ_0 that is complex homogeneous of degree 1 and satisfies $\delta\rho_0(\frac{1}{2i}\partial\overline{\partial}(F_{I(\alpha_0)})) = \frac{1}{2i}\partial\overline{\partial}F_0$, and so $f_\alpha \circ \rho_0$ has the same properties that f_α has, except that $\delta(f_\alpha \circ \rho_0)(\mu) = \frac{1}{2i}\partial\overline{\partial}F_0$.

Define $\psi_\alpha : M \to \mathcal{C}$ by $\psi_\alpha = f_\alpha \circ \rho_0 \circ \Pi$. Then $\psi_\alpha \in \widehat{\Sigma}_h$ for each α, and ψ_α depends holomorphically on α. Lemma 11.6 implies that $\psi_\alpha |_{\widehat{B}_n} \in \mathcal{S}^\infty$ for each α, and Proposition 8.1 tells us that $\psi_\alpha |_{\widehat{B}_n} = \hat{\sigma}_\alpha$ for some $\sigma_\alpha \in \mathcal{S}^\infty$. Because ψ_α maps M to the graph of $\partial F_{I(\alpha)}$, $\hat{\sigma}_\alpha$ maps $\widehat{B}_n$ into this same graph, and so we must have $F_{I(\alpha)} = F_0 \circ \sigma_\alpha^{-1}$. Thus $I(\alpha) = \sigma_\alpha(B_n)$, and since σ_α depends holomorphically on α, we have that $\alpha \mapsto I(\alpha)$ is special.

This finishes the proof of the "if" part of Theorem 11.5 in the case where $m = 1$. When $m > 1$ we cannot apply Theorem 8.11 in [S] directly, but it is not hard to make the changes needed for this situation. We omit the details.

12. A NATURAL ACTION ON $\widehat{\mathcal{R}}$

The conditions (9.5) on a mapping $\psi : \widehat{B}_n \to \mathcal{C}$ are clearly preserved if we compose ψ on the left by a holomorphic γ-symplectomorphism that also preserves Γ. This observation leads to a class of (locally-defined) transformations on $\widehat{\mathcal{R}}^\infty$ that we discuss in this section. We shall discuss in the next section the induced action on $\widehat{\mathcal{D}}^\infty$ and $\mathcal{D}^\infty$, and the relation between this action and the Kobayashi indicatrix.

We first spell out the kind of holomorphic symplectomorphisms that we are going to allow. Let $\mathcal{H}$ denote the set of holomorphic mappings h from some domain Ω in $\mathcal{C}$ (that depends on h) into $\mathcal{C}$ that satisfy the following properties.

First of all we require that Ω be a nonempty open set whose closure contains $(0,0)$. Set $\Omega^0 = \Omega \cup \{(0,0)\}$ and

$$\Omega^\epsilon = \{p \in \Omega^0 : \operatorname{dist}(p, \partial\Omega) \geq \epsilon \operatorname{dist}(p,(0,0))\} \tag{12.1}$$

for each $\epsilon > 0$. The conditions on h that we demand are:

(12.2) h is a biholomorphism of Ω onto its image, and h extends to a homeomorphism of Ω^0 to its image that sends $(0,0)$ to $(0,0)$;

(12.3) h is bilipschitz on every bounded subset of Ω^ϵ for each $\epsilon > 0$;

(12.4) $\delta h(\Gamma) = \Gamma,\ \delta h(\gamma) = \gamma \quad \text{on } \Omega.$

Notice that (12.2) and (12.3) imply that for every $\epsilon > 0$ and every bounded subset A of Ω^ϵ we have

$$C(A)^{-1} \operatorname{dist}(p,(0,0)) \leq \operatorname{dist}(h(p),(0,0)), \leq C(A) \operatorname{dist}(p,(0,0)), \tag{12.5}$$

$$|(dh(p))^{-1}| \leq C(A), \quad \text{and} \tag{12.6}$$

$$|\nabla^j h(p)| \leq C(A,j) \operatorname{dist}(p,(0,0))^{-j},\ j = 1,2,3,\ldots, \tag{12.7}$$

for all $p \in A$. For (12.7) it is helpful to employ the Cauchy formula for derivatives of holomorphic functions.

Fix $h \in \mathcal{H}$ with domain Ω. We define $\widehat{H}$, a mapping from a subset of $\widehat{\mathcal{R}}^\infty$ into $\widehat{\mathcal{R}}^\infty$, by $\widehat{H}(\psi) = h \circ \psi$ when ψ lies in

$$\text{domain } \widehat{H} = \{\psi \in \widehat{\mathcal{R}}^\infty : \psi(\widehat{B}_n) \subseteq \Omega^\epsilon \text{ for some } \epsilon > 0, \text{ and } \Pi \circ h \circ \psi \text{ is bilipschitz}\}. \tag{12.8}$$

Notice that $\Pi \circ h \circ \psi$ is bilipschitz if and only if $h(\psi(\widehat{B}_n))$ is the graph of a Lipschitz function over some domain in $\mathbf{C}^n$. Clearly $\widehat{H}(\psi) \in \widehat{\mathcal{R}}^\infty$ when ψ is in the domain of $\widehat{H}$, and it lies in $\widehat{\mathcal{R}}^\omega$ when ψ does.

Let us check that (12.8) defines an open subset of $\widehat{\mathcal{R}}^\infty$. Indeed, if $\psi, \psi' \in \widehat{\mathcal{R}}^\infty$, ψ lies in domain $\widehat{H}$, and if

$$\|d\psi' - d\psi\|_{L^\infty(\widehat{B}_n)} \leq \eta \tag{12.9}$$

for a sufficiently small η, then ψ' also lies in domain $\widehat{H}$. To see this we first observe that

$$|\psi(p) - \psi'(p)| \leq \eta \operatorname{dist}(p, (0,0)) \tag{12.10}$$

for $p \in \widehat{B}_n$, since $\psi((0,0)) = (0,0) = \psi'((0,0))$. Thus if $\psi(\widehat{B}_n) \subseteq \Omega^\epsilon$ for some $\epsilon > 0$, then $\psi'(\widehat{B}_n) \subseteq \Omega^\delta$ for some $\delta > 0$ as long as η is small enough. We also have

$$\|d(h \circ \psi) - d(h \circ \psi')\|_{L^\infty(\widehat{B}_n)} \leq C\eta$$

because of (12.9), (12.10), and (12.7) (with $j = 1, 2$). Hence $\Pi \circ h \circ \psi'$ is bilipschitz if $\Pi \circ h \circ \psi$ is and η is small enough, and so $\psi' \in$ domain $\widehat{H}$.

A very nice feature of $\widehat{H}$ is that it is holomorphic according to the conventions of Section 10. We can obviously extend $\widehat{H}$ to a holomorphic map of an open set in $C^\infty(\partial B, \mathcal{C})$ into $C^\infty(\partial B, \mathcal{C})$ defined by $\psi \mapsto h \circ \psi$. We can also build a holomorphic extension of $\widehat{H}$ to an open subset of $\widehat{C}_a^\infty$ into $\widehat{C}_a^\infty$ using this and Lemma 10.3.

The next result implies that there are plenty of elements of $\mathcal{H}$.

PROPOSITION 12.11. *Given $\psi_1, \psi_2 \in \widehat{\mathcal{R}}^\omega$ there is an $h \in \mathcal{H}$ such that $\psi_1 \in$ domain $\widehat{H}$ and $\widehat{H}(\psi_1) = \psi_2$.*

This follows immediately from Theorem 7.5, which implies that $\psi_2 \circ \psi_1^{-1}$ has a holomorphic extension to a neighborhood of $\psi_1(\widehat{B}_n)$ that lies in $\mathcal{H}$.

An important subset of $\mathcal{H}$ arises from biholomorphisms on $\mathbf{C}^n$. Let $f : U_1 \to U_2$ be a biholomorphism of U_1 onto U_2, where U_1, U_2 are domains in $\mathbf{C}^n$ that contain the origin, and suppose that $f(0) = 0$. Define $h : U_1 \times \mathbf{C}^n \to U_2 \times \mathbf{C}^n$ by (6.9), with g replaced by h. Then $h \in \mathcal{H}$, and $\widehat{H}$ can be described as follows. If $\rho \in \mathcal{R}^\infty$ and $\rho(\overline{B}_n) \subseteq U_1$, then $\hat{\rho}$ lies in the domain of $\widehat{H}$, and $\widehat{H}(\hat{\rho}) = (f \circ \rho)^\wedge$. Here $(f \circ \rho)^\wedge$ is defined as in (7.1); note that $f \circ \rho \in \mathcal{R}^\infty$.

When $n = 1$ all of $\mathcal{H}$ arises this way, and Proposition 12.11 can easily be derived directly from the classical Riemann mapping theorem.

Next we discuss the relationship between the action of $\mathcal{H}$ on $\widehat{\mathcal{R}}^\infty$ and $\widehat{S}$: $\widehat{\mathcal{R}}^\infty \to \widehat{\mathcal{S}}^\infty$.

PROPOSITION 12.12. *Suppose that $h \in \mathcal{H}$ and that $\psi_1, \psi_2 \in \widehat{\mathcal{R}}^\infty$ lie in the domain of $\widehat{H}$. If $\widehat{S}(\psi_1) = \widehat{S}(\psi_2)$, then $\widehat{S}(\widehat{H}(\psi_1)) = \widehat{S}(\widehat{H}(\psi_2))$.*

It is not hard to prove this directly, by showing that $\widehat{S}(\psi_1) = \widehat{S}(\psi_2)$ iff ψ_1 and ψ_2 are asymptotically the same at $(0,0)$ in a certain sense, and that $\widehat{H}$ preserves this. We won't provide the details of this argument, but instead we are going to do something more complicated that gives this proposition and a formula for $\widehat{S}(\widehat{H}(\psi))$.

Fix $h : \Omega \to \mathcal{C}$ in $\mathcal{H}$. We want to associate to h an element $\tilde{h} : \tilde{\Omega} \to \mathcal{C}$ in $\mathcal{H}$ that commutes with δ_λ and which approximates h well at $(0,0)$.

We would like to define $\tilde{h}$ by

$$\tilde{h} = \lim_{\lambda \to 0} \delta_\lambda^{-1} \circ h \circ \delta_\lambda, \qquad \lambda \in \mathbf{C} \setminus \{0\}. \tag{12.13}$$

Let $\tilde{\Omega}$ denote the set of p in $\mathcal{C}$ such that there exist $a, b > 0$ so that $\delta_\lambda(q) \in \Omega^0$ for all $q \in \mathcal{C}$, $|p - q| < a$, and all $\lambda \in \mathbf{C}$, $|\lambda| < b$, and such that (12.13) exists for all such q. Then we can and do define $\tilde{h}$ on $\tilde{\Omega}$ by (12.13), and we have that $\delta_\lambda(\tilde{\Omega}) = \tilde{\Omega}$ and $\tilde{h} \circ \delta_\lambda = \delta_\lambda \circ \tilde{h}$ for all $\lambda \in \mathbf{C} \setminus \{0\}$.

LEMMA 12.14. *$\tilde{h} : \tilde{\Omega} \to \mathcal{C}$ satisfies (12.2), (12.3), and (12.4).*

This is basically trivial. The main point is that the convergence in (12.13) is uniform on compact subsets of $\tilde{\Omega}$, because of (12.5) and the usual normal families arguments.

This lemma implies that $\tilde{h} \in \mathcal{H}$ if $\tilde{\Omega}$ is nonempty. This is true when the domain of $\widehat{H}$ is nonempty (which is the only case that we care about) because of the following result.

THEOREM 12.15. *$\widehat{S}$ maps the domain of $\widehat{H}$ into the domain of $\widehat{\tilde{H}}$, where $\widehat{\tilde{H}}$ is the analogue of $\widehat{H}$ for $\tilde{h}$, and*

$$\widehat{\tilde{H}} \circ \widehat{S} = \widehat{S} \circ \widehat{H}. \tag{12.16}$$

Notice that Proposition 12.12 is a special case of (12.16).

Before proving this theorem we record some remarks. Let $\tilde{\mathcal{H}}$ denote the subset of $\mathcal{H}$ of mappings which commute with δ_λ and whose domains are invariant under δ_λ for all $\lambda \in \mathbf{C} \setminus \{0\}$. Thus $\tilde{\mathcal{H}}$ induces an action on $\widehat{\mathcal{S}}^\infty$ as well as $\widehat{\mathcal{R}}^\infty$, since $\widehat{\tilde{H}}(\psi) \in \widehat{\mathcal{S}}^\infty$ if $\psi \in \widehat{\mathcal{S}}^\infty$ and $\tilde{h} \in \tilde{\mathcal{H}}$. The definition of $\tilde{H}$ can be simplified using the next lemma, which we prove after proving the theorem.

LEMMA 12.17. *If $g : \Omega \to \mathcal{C}$ is holomorphic, commutes with δ_λ, and satisfies $\delta g(\gamma) = \gamma$, then $\delta g(\Gamma) = \Gamma$.*

It follows from this that $\widetilde{\mathcal{H}}$ is essentially the same as what was called $\mathcal{H}\widehat{\Sigma}$ in Section 7 of [S], with $(\mathcal{C},\gamma)$ playing the role of $(\widehat{M},\widehat{\Omega})$ in [S]; the only significant difference is that we have added the homogeneity condition that the elements of $\widetilde{\mathcal{H}}$ commute with δ_λ. Because of this and Lemma 11.6, the action of $\widetilde{\mathcal{H}}$ on $\widehat{\mathcal{S}}^\infty$ is essentially just a special case of the action of $\mathcal{H}\widehat{\Sigma}$ on $\widehat{\Sigma}$ discussed in Section 7 in [S].

Let us now prove Theorem 12.15. Fix $\psi \in \widehat{\mathcal{R}}^\infty$ which lies in the domain of $\widehat{H}$, and let $\phi = S(\psi)$. We first observe that

$$\lim_{\lambda\to 0}\left(\sup_{p\in\widehat{B}_n} |\delta_\lambda^{-1}\circ\psi\circ\delta_\lambda(p) - \phi(p)|\,|p|^{-1}\right) = 0. \tag{12.18}$$

This works for any $\psi \in \widehat{C}_a^\infty$; it follows from a Cauchy integral representation for ψ like (4.7), but adapted to $\widehat{C}_a^\infty$ instead of C_a^∞. [Such a representation is given in the proof of Lemma 10.3.]

Choose $\epsilon > 0$ so that $\psi(\widehat{B}_n)$ is contained in $\Omega^{4\epsilon}$. From (12.18) it follows that there is a $t_0 > 0$ so that

$$\delta_\lambda(\phi(p)) \in \Omega^{3\epsilon} \text{ if } |\lambda| \le t_0 \text{ and } p \in \widehat{B}_n. \tag{12.19}$$

Next we want to show that (12.13) exists on $\phi(\widehat{B}_n)$. Set $\psi_1 = \widehat{H}(\psi)$, $\phi_1 = \widehat{S}(\psi_1)$. Then

$$\phi_1 = \lim_{\lambda\to 0}\delta_\lambda^{-1}\circ h\circ\psi\circ\delta_\lambda,$$

since $\psi_1 = h\circ\psi$. In particular we know that the right side exists, since $h\circ\psi \in \widehat{\mathcal{R}}^\infty \subseteq \widehat{C}_a^\infty$. It is not hard to check that

$$\delta_\lambda^{-1}\circ h\circ\psi\circ\delta_\lambda - \delta_\lambda^{-1}\circ h\circ\phi\circ\delta_\lambda$$

tends to zero uniformly on $\widehat{B}_n$ as $\lambda \to 0$. This follows from (12.18), (12.3), and the fact that $\psi(\widehat{B}_n) \subseteq \Omega^{4\epsilon}$. Thus

$$\phi_1 = \lim_{\lambda\to 0}\delta_\lambda^{-1}\circ h\circ\phi\circ\delta_\lambda = \lim_{\lambda\to 0}\delta_\lambda^{-1}\circ h\circ\delta_\lambda\circ\phi, \tag{12.20}$$

and so (12.13) does exist on the image of ϕ.

To show that $\phi(\widehat{B}_n) \subseteq \widetilde{\Omega}$ we must show that (12.13) converges on a δ_λ-invariant neighborhood of $\phi(\widehat{B}_n)$. Let Ω' denote the set of $p \in \mathcal{C}$ for which there exist a, $b > 0$ such that $\delta_\lambda(q) \in \Omega^0$ for all $q \in \mathcal{C}$ with $|p-q| < a$ and all $\lambda \in \mathbf{C}$ with $|\lambda| < b$. Let us check that (12.13) exists on the connected component U of Ω' that contains $\phi(\widehat{B}_n)\setminus\{0\}$.

If $\{\lambda_j\}$ is any sequence of complex numbers that goes to zero, then there is a subsequence $\{\xi_k\}$ such that $\delta_{\xi_k}^{-1}\circ h\circ\delta_{\xi_k}$ converges uniformly on compact subsets of U. This follows from (12.5) and a normal families argument. This limit does not depend on the sequence $\{\xi_k\}$: if $\{\eta_k\}$ is another sequence for which we have convergence, then

$$\lim_{k\to\infty}\delta_{\xi_k}^{-1}\circ h\circ\delta_{\xi_k} = \lim_{k\to\infty}\delta_{\eta_k}^{-1}\circ h\circ\delta_{\eta_k}$$

on $\phi(\widehat{B}_n)\setminus\{0\}$, and hence on all of U, because these limits are holomorphic on U, and because $\phi(\widehat{B}_n)\setminus\{(0,0)\}$ is totally real and has real dimension $2n$. Standard reasoning implies now that (12.13) exists on all of U.

Hence $U \subseteq \widetilde{\Omega}$. Using this and (12.19) it is easy to see that $\phi(\widehat{B}_n) \subseteq \widetilde{\Omega}^{2\epsilon}$. Since $\tilde{h} \circ \phi = \phi_1$, by (12.20), and $\phi_1 \in \widehat{S}^\infty$ we get that ϕ lies in the domain of $\widehat{\widehat{H}}$, and that $\widehat{\widehat{H}}(\phi) = \phi_1 = \widehat{S}(\widehat{H}(\psi))$. This proves Theorem 12.15, since ψ was arbitrary.

Let us prove Lemma 12.17. It suffices to show that $dg(V) = V$, where V is the holomorphic vector field on C given in z, ζ coordinates by $\Sigma\zeta_j \frac{\partial}{\partial\zeta_j}$, because of our observations in Section 8. (See (8.4).) On the other hand the hypothesis that $\delta_\lambda \circ g = g \circ \delta_\lambda$ implies that $dh(W_\lambda) = W_\lambda$ for each $\lambda \in \mathbf{C}$, where W_λ is the holomorphic vector field given by

$$\lambda \Sigma z_j \frac{\partial}{\partial z_j} + \overline{\lambda}\Sigma\zeta_j \frac{\partial}{\partial \zeta_j}.$$

Using this with $\lambda = 1, i$ gives $dg(V) = V$, as desired.

13. THE ACTION OF $\mathcal{H}$ ON DOMAINS IN $\mathbf{C}^n$

We observed in Section 9 that $\widehat{\mathcal{D}}^\infty$ can be viewed as $\widehat{\mathcal{R}}^\infty/\widehat{\mathcal{G}}^\infty$, where $\widehat{\mathcal{G}}^\infty$ acts on $\widehat{\mathcal{R}}^\infty$ by composition on the right. This commutes with the action of $\mathcal{H}$ on $\widehat{\mathcal{R}}^\infty$, which operates by composition on the left, and so $\mathcal{H}$ also gives an action on $\widehat{\mathcal{D}}^\infty$. To be specific, if $h \in \mathcal{H}$ and $\widehat{\mathcal{D}} \in \widehat{\mathcal{D}}^\infty$ are given, and if $\psi \in \widehat{\mathcal{R}}^\infty$ satisfies $\psi(\widehat{B}_n) = \widehat{D}$, then we say that $\widehat{D}$ lies in the domain of $\widehat{H}$ if $\hat{\psi}$ does, and we define $\widehat{H}(\widehat{D})$ to be the image of $\widehat{B}_n$ under $\widehat{H}(\psi) = h \circ \psi$. This is clearly independent of the particular choice of ψ.

We can also view this as an action on $\mathcal{D}^\infty$ instead of $\widehat{\mathcal{D}}^\infty$, using the correspondence between them. Given $h \in \mathcal{H}$, we denote the corresponding operation on $\mathcal{D}^\infty$ by H.

This action on $\mathcal{D}^\infty$ can be described more directly as follows. Fix $h : \Omega \to \mathcal{C}$ in $\mathcal{H}$ and $D \in \mathcal{D}^\infty$. Thus $\widehat{D} \in \widehat{\mathcal{D}}^\infty$ is the graph in $\mathcal{C}$ over D of ∂F, where $u = \frac{1}{2}\log F$ is the Green's function on D. If $\widehat{D} \subseteq \Omega^\epsilon$ for some $\epsilon > 0$ and $h(\widehat{D})$ is the graph of a Lipschitz function over some domain D', then D lies in the domain of H and $H(D) = D'$.

For this to work we don't really need D to be in $\mathcal{D}^\infty$; we only need that D contains the origin and that if u is the Green's function for D, then $F = e^{2u}$ is sufficiently well-behaved. In particular F should be reasonably smooth and strictly plurisubharmonic away from the origin. If the graph $\widehat{D}$ of ∂F over D in $\mathcal{C}$ is contained in Ω^ϵ for some $\epsilon > 0$, and $h(\widehat{D})$ is the graph of a Lipschitz function over some domain D', then in fact $h(\widehat{D})$ is the graph of $\partial F'$ over D', where $\frac{1}{2}\log F'$ is the Green's function for D', and we set $H(D) = D'$. [This is similar to the discussion in Section 7 in the vicinity of (7.14).]

There is a nice relationship between the action of $\mathcal{H}$ on $\mathcal{D}^\infty$ and the Kobayashi indicatrix.

THEOREM 13.1. *Let $h \in \mathcal{H}$ and $D \in \mathcal{D}^\infty$ be given, and suppose that D lies in the domain of H. Let $\tilde{h} \in \widetilde{\mathcal{H}}$ be as in Section 12, around* (12.13). *Then $K(D)$ lies in the domain of $\widetilde{H}$, and $K(H(D)) = \widetilde{H}(K(D))$.*

This follows from Theorems 11.2 and 12.15.

Notice that $\widetilde{\mathcal{H}}$ acts on $\mathcal{I}^\infty$ and not just $\mathcal{D}^\infty$. That is, if $\tilde{h} \in \widetilde{\mathcal{H}}$, $I \in \mathcal{I}$, and I lies in the domain of $\widetilde{H}$, then $\widetilde{H}(I)$ lies in $\mathcal{I}^\infty$, and not just $\mathcal{D}^\infty$. Thus Theorem 13.1 says that K intertwines the action of $\mathcal{H}$ on $\mathcal{D}^\infty$ and $\widetilde{\mathcal{H}}$ on $\mathcal{I}^\infty$.

The action of $\widetilde{\mathcal{H}}$ on $\mathcal{I}^\infty$ is a special case of something in [S]. To explain this we need some more notation. Let $\mathcal{N}_h^\infty$ denote the space of functions F on $\mathbf{C}^n$

such that

$$F(z) > 0 \text{ unless } z = 0, \ \ F(\lambda z) = |\lambda|^2 F(z)$$

for all $z \in \mathbf{C}^n$, and F is C^∞ and strictly plurisubharmonic except at the origin. Thus $I \mapsto F_I$ is a bijection between $\mathcal{I}^\infty$ and $\mathcal{N}_h^\infty$. The action of $\widetilde{\mathcal{H}}$ on $\mathcal{I}^\infty$ produces an action on $\mathcal{N}_h^\infty$, which is an example of the kind of action discussed in Sections 7 and 8 in [S], modulo some minor modifications (such as keeping track of the homogeneity) which are discussed in part in Section 9 in [S]. This is closely related to the comments just after the statement of Lemma 12.17.

14. RIEMANNIAN GEOMETRY ON $\mathcal{D}^\infty$; PRELIMINARY DISCUSSION

This and the next sections will be dealing with a certain Riemannian structure on $\mathcal{D}^\infty$ and related topics.

Since we don't have any sort of reasonable manifold structure on $\mathcal{D}^\infty$ we have to be careful about what a "Riemannian structure" means. Most of what we do will have at least some kind of formal meaning on $\mathcal{D}^\infty$ which could be made precise in circumstances where the analysis is well under control, but we shall often simply restrict ourselves to the space $\mathcal{D}^\infty_{co}$ of smooth, strongly convex domains, where the analysis is reasonably well under control, because of the work of Lempert. (See Section 20 for a different approach.)

The sort of Riemannian metrics that we shall consider are merely what some authors would call weak Riemannian metrics. That is, these metrics are defined in terms of bilinear forms that are positive-definite but for which the associated norm does not determine the topology of the underlying vector space. A typical example of such a bilinear form is the L^2 inner product on a space of smooth functions.

The Riemannian metric on $\mathcal{D}^\infty$ that we shall consider is given in terms of an L^2 inner product of variations of the Green's function. A more precise but formal description goes as follows. Fix $D \in \mathcal{D}^\infty$, and let $\delta_1 D$, $\delta_2 D$ denote two first-order infinitesimal variations of D. These can be represented by a pair of sections of the normal bundle of ∂D, for example. Let u denote the Green's function of u, and let $\delta_1 u$, $\delta_2 u$ denote the corresponding first-order variations of the Green's functions. Then we want to take

$$g_D(\delta_1 D, \delta_2 D) = (-1)^n \int_{\partial D} (\delta_1 u)(\delta_2 u) d^c u \wedge \left(\frac{1}{2i}\partial\overline{\partial} u\right)^{n-1}, \tag{14.1}$$

where $d^c = i(\overline{\partial} - \partial)$, as usual.

Note that $(-1)^n d^c u \wedge \left(\frac{1}{2i}\partial\overline{\partial}u\right)^{n-1}$ defines a positive measure on ∂D when the latter is given its usual orientation. This is easily checked by using a $\rho : B_n \to D$ in $\mathcal{R}^\infty$ to reduce to the case of $B_n = D$, where it gives a positive constant times Lebesgue measure.

For this to be well defined the behavior of the variations of the Green's functions up to the boundary of D needs to be controlled. When $n = 1$ this is easily achieved with enough regularity assumptions, and when $n > 1$ you can do this for $D \in \mathcal{D}^\infty_{co}$ using the work of Lempert.

This Riemannian metric has a number of nice features. By defining it in terms of the Green's function it is reasonable to hope that the geometry on $\mathcal{D}^\infty$ will respect somehow the complex analysis back on the domains. In particular (14.1) is invariant under biholomorphisms in the following sense. If f is a biholomorphism of a neighborhood of $\overline{D}$ to an open set in $\mathbf{C}^n$, and if $f(0) = 0$, then f induces a map from a neighborhood of D in $\mathcal{D}^\infty$ onto a neighborhood of $f(D)$ in $\mathcal{D}^\infty$. It is easy to see that this map preserves (14.1), because of the invariance properties of the Green's function. In fact we'll see that (14.1) is preserved by the action of $\mathcal{H}$ described in the preceeding section.

In turns out that (14.1) interacts well with $\mathcal{I}^\infty$ and $K : \mathcal{D}^\infty \to \mathcal{I}^\infty$. We shall see that $\mathcal{I}^\infty$ is a totally geodesic subspace of $\mathcal{D}^\infty$, and that K is a Riemannian submersion, at least formally, and that these statements are true in a reasonable and precise sense in the strongly convex case.

The metric on $\mathcal{I}^\infty$ induced from $\mathcal{D}^\infty$ is very special. In this case there is no problem with making sense of (14.1), because the Green's function u_I on $I \in \mathcal{I}^\infty$ is given by $u_I = \frac{1}{2}\log F_I$, F_I as in Section 3. It is better to express (14.1) in terms of F_I, and to view it as a metric on the space $\mathcal{N}_h^\infty$ (defined at the end of the preceeding section). The resulting Riemannian metric on $\mathcal{N}_h^\infty$ is the same as the kind discussed in Section 4 of [S], except for the usual modifications needed to account for the noncompactness of $\mathbf{C}^n$ and the required homogeneity of the functions.

From [S] we know that $\mathcal{N}_h^\infty$, equipped with this metric, has parallel curvature tensor and is, in a reasonable sense, a locally symmetric space. (The relationship between these two properties is not as cozy in infinite dimensions as it is in finite dimensions.) Its geodesics are characterized by an equation which is a special case of HCMA, and in particular they give rise to a certain class of special mappings (in the sense of Section 11). In the convex case they can also be characterized in terms of the complex method of interpolation of Banach spaces, by a result of Rochberg [R].

One aspect of (14.1) that is somewhat arbitrary is the choice of integrating over ∂D instead of D (with $F = e^{2u}$ to some power, say, inserted into the integral to kill the singularity at the origin). For many purposes this doesn't matter, but for others (like computing the equation for geodesics) it is better to integrate over ∂D.

15. SOME BASIC FACTS AND DEFINITIONS CONCERNING THE METRIC ON $\mathcal{D}^{\infty}_{co}$

Throughout this section and the next four we restrict ourselves to strongly convex domains. This will not play a role in the formal computations, but it will allow us to make rigorous and precise statements, using the work of Lempert.

For the most part we shall not discuss the geometry of $\mathcal{D}^{\infty}_{co}$ in the conventional manner, in terms of tangent spaces and bilinear forms, but rather in terms of the behavior of the energy of curves in $\mathcal{D}^{\infty}_{co}$.

Let D_t, $0 \le t \le 1$, be a curve of domains in $\mathcal{D}^{\infty}_{co}$. It is convenient to assume that the dependence on t is C^{∞}. This notion can easily be made precise by parameterizing the set of domains near a given domain by sections of the normal bundle of the boundary of the given domain that are close to the zero section, using the fact that there is a neighborhood of the boundary of any smooth domain that is diffeomorphic to a neighborhood of the zero section in the normal bundle.

Set $I_t = K(D_t)$, and let $\Psi_t : \overline{I}_t \to \overline{D}_t$ be as in (5.1). Let F_{I_t} be the function associated to I_t as in Section 3. Define $\Psi^o_t : \overline{B}_n \to \overline{D}_t$ by

$$\Psi^o_t(z) = \Psi_t\left(\frac{|z|}{(F_{I_t}(z))^{\frac{1}{2}}} z\right). \tag{15.1}$$

Notice that Ψ^o_t does not lie in $\mathcal{R}^{\infty}$, unless I_t is a dilate of B_n, because it won't satisfy (1.4). On the other hand, it does lie in C^{∞}_a, and for each $z \in \partial B_n$ the mapping $\lambda \mapsto \Psi^o_t(\lambda z)$ is an extremal mapping of Δ into D_t in the sense described in Section 5. The following will provide us with control on the dependence in t of the Green's function on D_t.

THEOREM 15.2. *(Lempert [L1].) $t \mapsto \Psi^o_t$ is a C^{∞} map of $[0, 1]$ into C^{∞}_a.*

Let u_t denote the Green's function on D_t. This is inconsistent with our earlier use of u_0 for the Green's function on B_n; we now denote that by u^o.

It is easy to see that

$$u_t \circ \Psi^o_t = u^o. \tag{15.3}$$

This uses Lemma 3.5. From here it follows that $\dot{u}_t \circ \Psi^o_t(z)$ is smooth away from $z = 0$, where the dot denotes the derivative in t. Hence the energy of D_t defined according to (14.1) — i.e.,

$$E = E(D_t) = (-1)^n \int_0^1 \int_{\partial D_t} (\dot{u}_t)^2 d^c u_t \wedge \left(\frac{1}{2i}\partial\overline{\partial} u_t\right)^{n-1} dt \tag{15.4}$$

— makes sense and is finite.

To understand this energy functional better we must understand $\dot{u}_t$ better. We should find a good way to represent infinitesimal deformations of a domain, and a good formula for the corresponding first-order variation of the Green's function. For this purpose, which will occupy us for most of the remainder of this section, it is more convenient for us to fix a particular domain $D \in \mathcal{D}^\infty_{co}$ and work with it.

A first-order variation of D is represented by a vector field along ∂D. Two vector fields represent the same deformation if they differ by a tangent vector field. If we choose a (real) line bundle on ∂D that is transverse to ∂D, then variations can be represented uniquely by its sections. In our case there is a good choice for this line bundle.

There are a number of subbundles of

$$T = T(D) = \text{ the tangent bundle to } \mathbf{C}^n \text{ restricted to } \overline{D} \setminus \{0\}$$

that will be relevant for us, some of which we have used before. Let $\rho \in \mathcal{R}^\infty$ be such that $\rho(B_n) = D$, and let $T^1 = T^1(D)$, $T^2 = T^2(D)$ denote the subbundles of T obtained by pushing S^1, S^2 (on B_n) forward using $d\rho$. Let $T^l = T^l(D)$ denote the subbundle of T such that T^l_z is the tangent space of the level surface of the Green's function u on D through z. Thus T^2 is the maximal complex subbundle of T^l. Let $T^e = T^e(D)$ denote the real line bundle such that $J(T^e) = T^1 \cap T^l$, where J denotes the standard complex structure on T, and let $T^e(\partial D)$ denote the restriction of T^e to ∂D. Notice that $T^e(\partial D)$ is transverse to ∂D. ("e" stands for "exit"; "n" for "normal" is already occupied, and not completely appropriate.) Clearly T is the direct sum of T^1 and T^2, as well as T^e and T^l.

According to our previous remark we can represent infinitesimal variations of D by sections of $T^e(\partial D)$. It turns out to be more convenient to use a different but related representation which comes from the next lemma. Let $\Psi^o : \overline{B}_n \to \overline{D}$ be as above.

Lemma 15.5. *Let X be a smooth section of $T^e(\partial D)$. Then there is a unique vector field Y on $\overline{D}$ with the following properties:*

(15.6) $X - Y \big|_{\partial D}$ *is tangent to* ∂D;

(15.7) $Y(z) \in T^1_z$ *for* $z \neq 0$, *and* $Y(0) = 0$;

(15.8) $Y(\Psi^o(\lambda z))$ *is holomorphic in* $\lambda \in \Delta$ *for all* $z \in \partial B_n$;

(15.9) *for each* $z \in \partial B_n$ *there is a real number* $a(z)$ *such that*

$$\frac{\partial}{\partial \lambda}\Big|_{\lambda=0} (Y(\Psi^o(\lambda z))) = a(z) \frac{\partial}{\partial \lambda}\Big|_{\lambda=0} \Psi^o(\lambda z).$$

Moreover, Y is smooth away from the origin.

In (15.9) the point is that $a(z)$ is real; that such an $a(z) \in \mathbf{C}$ exists follows from (15.7) and (15.8).

This lemma is essentially harmonic conjugation on the unit disk dressed up in the notation of our situation. Because X is a section of $T^e(\partial D)$ there is a real-valued function f on ∂B_n such that

$$X(\Psi^o(z)) = f(z)\frac{\partial}{\partial\lambda}\Big|_{\lambda=1}(\Psi^o(\lambda z)). \tag{15.10}$$

This comes from chasing definitions. There is a unique complex-valued function g on $\overline{B}_n \setminus \{0\}$ such that $\operatorname{Re} g\,\big|_{\partial B_n} = f$ and $g(\lambda z)$ extends to a holomorphic function of $\lambda \in \Delta$ which is real at $\lambda = 0$ for all $z \in \partial B_n$. We take Y to be given by

$$Y(\Psi^o(z)) = g(z)\frac{\partial}{\partial\lambda}\Big|_{\lambda=1}(\Psi^o(\lambda z)). \tag{15.11}$$

It is not hard to check that Y has the desired properties, and that the uniqueness of Y follows from the uniqueness of g.

Let $\mathcal{T} = \mathcal{T}(D)$ denote the space of vector fields Y that satisfy (15.7), (15.8), and (15.9) above. We view the elements of $\mathcal{T}$ as representing the first-order infinitesimal deformations of D, so that $\mathcal{T}(D)$ represents the tangent space of $\mathcal{D}^\infty_{co}$ at D.

The next lemma will be helpful in producing elements of $\mathcal{T}$. Given $z \in \overline{D}\setminus\{0\}$, let $P_z : \mathbf{C}^n \to T^1_z$ denote the projection onto T^1_z with kernel T^2_z.

Lemma 15.12. *Fix $z \in \partial B_n$. Then $P_{\Psi^o(\lambda z)}$ is holomorphic in $\lambda \in \Delta \setminus \{0\}$ and has a holomorphic extension across $\lambda = 0$.*

In other words, if $f : \Delta \to D$ is one of the extremal mappings (as discussed in Section 5), then $P_{f(\lambda)}$ is holomorphic in λ, even at $\lambda = 0$.

For $D \in \mathcal{D}^\infty_{co}$ this is proved in [L1]. We include the proof here for completeness and to make clear its validity for $D \in \mathcal{D}^\infty$ (or for $D = \rho(B_n)$ for any ρ that satisfies (1.1)–(1.4)).

Let u be the Green's function on D. Let us first check that

$$P_{f(\lambda)}(v) = 2\lambda f'(\lambda)\partial u_{f(\lambda)}(v) \tag{15.13}$$

for any $v \in \mathbf{C}^n$. Clearly the right side vanishes when $v \in T^2_{f(\lambda)}$, and so it remains to verify that the right side equals v when $v \in T^1_{f(\lambda)}$. We may as well take $v = f'(\lambda)$. Because $u(f(\lambda)) = \log|\lambda|$, we have that

$$\frac{\partial}{\partial\lambda}(\log|\lambda|) = \frac{\partial}{\partial\lambda}(u(f(\lambda))) = \partial u_{f(\lambda)}(f'(\lambda)).$$

It is easy to check that the left side is $(2\lambda)^{-1}$, and (15.13) follows easily.

Next we check that $\partial u_{f(\lambda)}$ is holomorphic in λ when $\lambda \neq 0$. This is provided by the proof of $(\partial\overline{\partial}u)^n \equiv 0$, but we give another (slightly stupid) argument now. Because $u(f(\lambda))$ is harmonic in λ we have that

$$\sum_{j,k}(\partial_j\overline{\partial}_k u)(f(\lambda))\; f_j'(\lambda)\overline{f_k'(\lambda)} = 0.$$

On the other hand, we know that u is plurisubharmonic, so that $\partial_j\overline{\partial}_k u$ must be a nonnegative matrix, and hence

$$\sum_k (\partial_j\overline{\partial}_k u)(f(\lambda))\,\overline{f_k'(\lambda)} = 0$$

for all j. Thus $\partial u_{f(\lambda)}$ is holomorphic in $\lambda \in \Delta \setminus \{0\}$.

To get the holomorphic extension across $\lambda = 0$ it suffices to show that $|\partial u_z| = O(|z|^{-1})$ as $z \to 0$. One way to see this is to use the fact that $\delta\rho(\partial u) = \partial u^o$ when $\rho : B_n \to D$ satisfies (1.1)–(1.4). This completes the proof of Lemma 15.12.

Let us come back now to our curve D_t in $\mathcal{D}^\infty_{co}$, and let's look at how $\frac{d}{dt}D_t$ is represented by an element of $\mathcal{T}(D_t)$. Let P^t be as above, but for D_t instead of D, so that P^t maps $T(D_t)$ onto $T^1(D_t)$ and has kernel $T^2(D_t)$.

LEMMA 15.14. *$Y_t = P^t\{\dot{\Psi}^o_t \circ (\Psi^o_t)^{-1}\}$ lies in $\mathcal{T}(D_t)$ and represents $\frac{d}{dt}D_t$.*

Clearly Y_t represents $\frac{d}{dt}D_t$, since $Y_t\,|_{\partial D} - \dot{\Psi}^o_t \circ (\Psi^o_t)^{-1}\,|_{\partial D}$ is tangent to ∂D (and even lives in $T^2(D_t)$), and because $\Psi^o_t(B_n) = D_t$. To show that $Y_t \in \mathcal{T}(D_t)$ we must check that Y_t satisfies (15.7), (15.8), and (15.9) on D_t. This is easily accomplished, using Lemma 15.12 and the various properties of Ψ^o_t. For (15.9) it is good to observe that

$$\text{(15.15)} \qquad \frac{\partial}{\partial\lambda}\bigg|_{\lambda=0} \Psi^o_t(\lambda z) = \frac{|z|}{(F_{I_t}(z))^{\frac{1}{2}}}\, z,$$

and so $\frac{\partial}{\partial t}$ of this is clearly a real multiple of itself.

Using (15.3) we obtain the following formula for $\dot{u}_t$ in terms of Y_t:

$$\text{(15.16)} \qquad \dot{u}_t = -Y_t(u_t).$$

Here $Y_t(u_t)$ denotes the action of Y_t, as a real vector field, on the function u_t. [This can be expressed in terms of the complex components of Y_t using a formula like (3.6).]

There is another way to derive (15.16) which is more awkward to justify but which fits somewhat better into the general scheme of things. It is not hard to see that (15.16) ought to be true on ∂D_t, because $u_s = 0$ on ∂D_s for all s and Y_t represents $\frac{d}{dt}D_t$. Both sides of (15.16) satisfy the linearization of HCMA about u_t away from $z = 0$, which in this case just means that both sides are harmonic on the image of all the extremal maps of Δ into D_t. These two facts, plus some control on the two sides of (15.16) near the origin (e.g., boundedness) allow you to obtain the equality.

We can reformulate (15.16) as follows. If D is any element of $\mathcal{D}^\infty_{co}$ with Green's function u, and if $Y \in \mathcal{T}(D)$ is given, then $-Y(u)$ gives the first-order variation of u corresponding to the infinitesimal deformation of D represented by Y. Thus we can rewrite (14.1) as

$$\text{(15.17)} \qquad g_D(Y_1, Y_2) = (-1)^n \int_{\partial D} Y_1(u) Y_2(u)\, d^c u \wedge \left(\frac{1}{2i}\partial\overline{\partial}u\right)^{n-1}$$

for all $Y_1, Y_2 \in \mathcal{T}(D)$.

Notice, incidentally, that Y is determined by $Y(u)$ if $Y \in \mathcal{T}(D)$. This is because $Y(u) = 0$ implies that $Y \mid_{\partial D}$ is tangent to ∂D. If Y is given by (15.11), then it can be checked that

$$(Y(u)) \circ \Psi^o = \mathrm{Re}(g). \tag{15.18}$$

16. THE METRIC ON $\mathcal{D}_{co}^{\infty}$, CIRCLED DOMAINS, AND THE KOBAYASHI INDICATRIX

In this section we are going to look at the behavior of $\mathcal{I}_{co}^{\infty} = \mathcal{I}^{\infty} \cap \mathcal{D}_{co}^{\infty}$ as a submanifold of $\mathcal{D}_{co}^{\infty}$, and also the behavior of $K : \mathcal{D}_{co}^{\infty} \to \mathcal{I}_{co}^{\infty}$, from the point of view of the Riemannian structure on $\mathcal{D}_{co}^{\infty}$.

We shall follow closely the notations of the preceeding section. In particular we let D_t be a curve in $\mathcal{D}_{co}^{\infty}$, $I_t = K(D_t)$, etc. One of the things we want to do is compute $\frac{d}{dt} I_t$ in terms of $\frac{d}{dt} D_t$, which amounts to computing the differential of K. For this we need some more notation.

Let $D \in \mathcal{D}_{co}^{\infty}$ be given, and let $\mathcal{T} = \mathcal{T}(D)$ be as before. There is a natural splitting of $\mathcal{T}$ into $\mathcal{T}_0 + \mathcal{T}_1$, where $Y \in \mathcal{T}$ lies in $\mathcal{T}_0$ if

$$\frac{\partial}{\partial \lambda}\bigg|_{\lambda=0} Y(\Psi^o(\lambda z)) = 0 \tag{16.1}$$

for all $z \in \partial B_n$, and Y lies in $\mathcal{T}_1$ if there is a real-valued function f on $\overline{B}_n \setminus \{0\}$ such that

$$Y(\Psi^o(z)) = f(z)\left(\frac{\partial}{\partial \lambda}\bigg|_{\lambda=1} \Psi^o(\lambda z)\right) \tag{16.2}$$

for all $z \in B_n \setminus \{0\}$, and if

$$f(z) = f(\lambda z) \tag{16.3}$$

for all $z \in B_n \setminus \{0\}$, $\lambda \in \Delta \setminus \{0\}$. Notice that $\mathcal{T}_0$ is a complex vector space.

Now let's return to the problem of computing $\frac{\partial}{\partial t} I_t$. Let Y_t be as in Lemma 15.14, so that Y_t represents $\frac{\partial}{\partial t} D_t$. Let $Y_{0,t}$ and $Y_{1,t}$ be the $\mathcal{T}_0(D_t)$ and $\mathcal{T}_1(D_t)$ parts of Y_t.

PROPOSITION 16.4. *Notation as above. Let X_t be the vector field on I_t defined by $d\Psi_t(X_t) = Y_{1,t}$. Then $X_t \in \mathcal{T}_1(I_t)$ and X_t represents $\frac{\partial}{\partial t} I_t$.*

Before giving the proof of this, which is pretty easy, let us make some remarks about what it means.

This proposition can be viewed as providing a formula for the differential dK of $K : \mathcal{D}_{co}^{\infty} \to \mathcal{I}_{co}^{\infty}$. Given $D \in \mathcal{D}_{co}^{\infty}$, dK_D is a linear map from $\mathcal{T}(D)$ to $\mathcal{T}(K(D))$ which vanishes on $\mathcal{T}_0(D)$ and sends $\mathcal{T}_1(D)$ onto $\mathcal{T}_1(K(D))$ in the obvious way. Namely, if $Y \in \mathcal{T}_1(D)$, then $dK_D(Y) = d(\Psi^{-1})(Y)$.

A consequence of this is that $\mathcal{T}_0(D)$ represents the tangent space at D to the "submanifold" of $\mathcal{D}^\infty_{co}$ consisting of elements of $\mathcal{D}^\infty_{co}$ having the same Kobayashi indicatrix. "Submanifold" is in quotes here because it is not clear in what sense this subset of $\mathcal{D}^\infty_{co}$ is a submanifold; however, Proposition 16.4 does provide the preceeding statement with some rigorous content.

Similarly, if $D \in \mathcal{I}^\infty_{co}$, then $\mathcal{T}_1(D)$ represents the tangent space at D to $\mathcal{I}^\infty_{co}$. This can be checked directly.

Let us prove the proposition. From (15.1) we have

$$\dot{\Psi}^o_t(z) = \dot{\Psi}_t\left(\frac{|z|}{(F_{I_t}(z))^{\frac{1}{2}}}z\right) + \frac{\partial}{\partial s}\bigg|_{s=t}\left\{\Psi_t\left(\frac{|z|}{(F_{I_s}(z))^{\frac{1}{2}}}z\right)\right\}. \tag{16.5}$$

Let $W_{0,t}$ and $W_{1,t}$ denote the vector fields on D_t such that $W_{0,t}\circ\Psi^o_t$ and $W_{1,t}\circ\Psi^o_t$ are the first and second terms on the right side of (16.5), respectively. Then $W_{0,t}$ and $W_{1,t}$ satisfy the following two important properties:

$$\frac{\partial}{\partial\lambda}\bigg|_{\lambda=0}(W_{0,t}\circ\Psi^o_t(\lambda z)) = 0, \tag{16.6}$$

because $\frac{\partial}{\partial\lambda}\big|_{\lambda=0}(\Psi_t(\lambda z)) = z$ for all t (by (5.1)); and

(16.7)
$$W_{1,t}(\Psi^o_t(z)) = \left\{\left(\frac{|z|}{(F_{I_t}(z))^{\frac{1}{2}}}\right)^{-1}\left(\frac{\partial}{\partial t}\left(\frac{|z|}{(F_{I_t}(z))^{\frac{1}{2}}}\right)\right)\right\}\left(\frac{\partial}{\partial\lambda}\bigg|_{\lambda=1}\Psi^o_t(\lambda z)\right),$$

which can easily be computed.

From (16.7) we get that $W_{1,t} \in \mathcal{T}_1(D_t)$, while (16.6) implies that $P^t(W_{0,t}) \in \mathcal{T}_0(D_t)$. Hence

$$Y_{0,t} = P^t(W_{0,t}),\ Y_{1,t} = W_{1,t}.$$

Let us calculate X_t. Using the representation of $W_{1,t}\circ\Psi^o_t$ as the last term in (16.5) and the definition of X_t we obtain that

$$X_t\left(\frac{|z|}{(F_{I_t}(z))^{\frac{1}{2}}}z\right) = \frac{\partial}{\partial t}\left(\frac{|z|}{(F_{I_t}(z))^{\frac{1}{2}}}\right)z$$

for all $z \in B_n$. Clearly this lies in $\mathcal{T}_1(I_t)$, and the fact that this represents $\frac{d}{dt}I_t$ amounts to a chasing of definitions. For example, if we let $\Psi^o_{I_t} : B_n \to I_t$ be defined like Ψ_t was for D_t, then

$$\Psi^o_{I_t}(z) = \frac{|z|}{(F_{I_t}(z))^{\frac{1}{2}}}z, \tag{16.8}$$

and so Lemma 15.4 implies that X_t represents $\frac{d}{dt}I_t$.

The next result says that $K : \mathcal{D}^\infty_{co} \to \mathcal{I}^\infty_{co}$ is a Riemannian submersion.

PROPOSITION 16.9. *Let $D \in \mathcal{D}^\infty_{co}$ be given, and set $I = K(D)$. Then $\mathcal{T}_0(D)$ and $\mathcal{T}_1(D)$ are orthogonal with respect to g_D defined by (15.17), and the map $Y \mapsto X = d(\Psi^{-1})(Y)$ sends $\mathcal{T}_1(D)$ isometrically onto $\mathcal{T}_1(I)$, when $\mathcal{T}_1(D)$, $\mathcal{T}_1(I)$ are equipped with g_D, g_I given by (15.17).*

To prove this we use the following.

PROPOSITION 16.10. *Suppose that* $D \in \mathcal{D}^\infty_{co}$ *and* $\rho \in \mathcal{R}^\infty$ *satisfies* $\rho(B_n) = D$. *Then* $d\rho$ *maps* $\mathcal{T}(B_n)$ *isometrically (with respect to* g_{B_n}, g_D *in (15.17)) onto* $\mathcal{T}(D)$, *and it sends* $\mathcal{T}_0(B_n)$ *onto* $\mathcal{T}_0(D)$, $\mathcal{T}_1(B_n)$ *onto* $\mathcal{T}_1(D)$.

This is an easy exercise. The point is that ρ does all the right things to extremal mappings, Green's functions, etc.

Most of Proposition 16.9 can be derived directly from Proposition 16.10 and the fact that $\Psi : I \to D$ can be written as $\Psi = (\Psi \circ \sigma) \circ \sigma^{-1}$, where $\sigma : B_n \to I$ and $\Psi \circ \sigma : B_n \to D$ lie in $\mathcal{R}^\infty$, by Theorems 5.2 and 3.4. It remains to check that $\mathcal{T}_0(D)$ and $\mathcal{T}_1(D)$ are orthogonal. For this we can reduce to $D = B_n$, because of Proposition 16.10.

This is very easy, but let's write it down anyway. If $Y_0 \in \mathcal{T}_0(B_n)$, $Y_1 \in \mathcal{T}_1(B_n)$, then by definitions we have $Y_0(z) = g_0(z)z$, $Y_1(z) = g_1(z)z$, where $g_0(z)$ is complex-valued, $g_0(\lambda z)$ is holomorphic in $\lambda \in \Delta$ for all $z \in \partial B_n$, and $g_0(0) = 0$, while g_1 is real-valued, $g_1(z)$ is defined only for $z \in B_n \setminus \{0\}$, and $g_1(\lambda z) = g_1(z)$ for all $z \in B_n \setminus \{0\}$, $\lambda \in \Delta \setminus \{0\}$. Thus

$$Y_0(u^0)(z) = 2\,\mathrm{Re} \sum g_0(z) z_j \partial_j u^0(z) = \mathrm{Re}(g_0(z))$$

since $u^0(z) = \log|z|$, and $Y_1(u^0)(z) = g_1(z)$. Hence

$$g_{B_n}(Y_0, Y_1) = c_n \int_{\partial B_n} \mathrm{Re}(g_0(z)) g_1(z)\, dz,$$

where dz denotes Lebesgue measure, and c_n is some constant that depends only on n. We can rewrite this as

$$\frac{c_n}{2\pi} \int_0^{2\pi} \int_{\partial B_n} \mathrm{Re}(g_0(e^{i\theta} z)) g_1(e^{i\theta} z)\, dz\, d\theta = \frac{c_n}{2\pi} \int_{\partial B_n} g_1(z)\, \mathrm{Re}\left(\int_0^{2\pi} g_0(e^{i\theta} z)\, d\theta \right) dz.$$

Our assumptions on g_0 imply that the inner integral vanishes, as desired.

Combining Propositions 16.9 and 16.4 with (15.9) and (15.16) we get the following.

PROPOSITION 16.11. *Let* D_t, I_t, Y_t, *and* $Y_{0,t}$ *be as before. Then*

$$E(D_t) = E(I_t) + \int_0^1 g_{D_t}(Y_{0,t}, Y_{0,t})\, dt. \tag{16.12}$$

In particular, $E(D_t) \geq E(I_t)$ *with equality if and only if* $Y_t \in \mathcal{T}_1(D_t)$ *for each* t.

Thus $\mathcal{I}^\infty_{co}$ is totally geodesic in $\mathcal{D}^\infty_{co}$.

17. THE RIEMANNIAN METRIC AND THE ACTION OF $\mathcal{H}$

In this section we shall show that $\mathcal{H}$ acts by isometries on $\mathcal{D}^\infty_{co}$, that it preserves the splitting of the tangent bundle of $\mathcal{D}^\infty_{co}$ into $\mathcal{T}_0$ and $\mathcal{T}_1$, and we'll discuss some consequences of these facts.

Fix $h : \Omega \mapsto \mathcal{C}$ in $\mathcal{H}$, and let D_t, $0 \le t \le 1$, be a smooth curve in $\mathcal{D}^\infty_{co}$ that is contained in the domain of H, as discussed in Section 13. We do not need to require the convexity of $H(D_t)$; we have never needed convexity for anything but controlling the underlying analysis, and in this case we can do that by reducing to corresponding facts for D_t.

As usual we let u_t be the Green's function for D_t, $F_t = \exp(2u_t)$, and we let $\widehat{D}_t$ be the graph in $\mathcal{C}$ over D_t of ∂F_t. Since the curve D_t is in the domain of H there is an $\epsilon > 0$ so that $\widehat{D}_t \subseteq \Omega^\epsilon$ for each t, and $h(\widehat{D}_t)$ is the graph of a Lipschitz function over some domain $D'_t \in \mathcal{D}^\infty$. Let u'_t, F'_t, $\widehat{D}'_t$ be as usual, so that $h(\widehat{D}_t) = \widehat{D}'_t$.

Define $\tau_t : D_t \to \widehat{D}_t$ by $\tau_t(z) = (z, \partial F_t(z))$, and similarly for $\tau'_t : D'_t \to \widehat{D}'_t$. Let ρ_t be some smooth curve in $\mathcal{R}^\infty$ such that $\rho_t(B_n) = D_t$. [We can take $\rho_t = \Psi_t \circ \sigma_t$, as in Theorem 5.2, where σ_t is a smooth curve in $\mathcal{S}^\infty$ with $\sigma_t(B_n) = K(D_t)$. The existence of such a smooth curve σ_t can be obtained as in the proof of Theorem 3.4.] Set $\rho'_t = H(\rho_t)$, so that $\hat{\rho}'_t = h \circ \hat{\rho}_t$, and ρ'_t is a smooth curve in $\mathcal{R}^\infty$ with $\rho'_t(B_n) = D'_t$.

Define θ_t by $\theta_t = \rho'_t \circ \rho_t^{-1}$, so that θ_t maps D_t to D'_t and

$$\tau'_t \circ \theta_t = h \circ \tau_t. \tag{17.1}$$

Notice that the ρ_t's are not uniquely determined by the D_t's, because you can compose the ρ_t's on the right by elements of $\mathcal{G}^\infty$, but that θ_t is determined by the D_t's and h.

PROPOSITION 17.2. *Suppose that $Y_t \in \mathcal{T}(D_t)$ represents $\frac{d}{dt}D_t$. Then $Y'_t = d\theta_t(Y_t)$ lies in $\mathcal{T}(D'_t)$ and represents $\frac{d}{dt}D_t$.*

Before proving this we discuss its implications.

Let us reformulate the content of this proposition in terms of the differential of H. It says that if $D \in \mathcal{D}^\infty_{co}$ lies in the domain of H, then $dH_D : \mathcal{T}(D) \to \mathcal{T}(H(D))$ is given by $Y \mapsto d\theta(Y)$, where θ maps D onto $H(D)$ and is of the form $\rho' \circ \rho^{-1}$ for some $\rho, \rho' \in \mathcal{R}^\infty$. From Proposition 16.10 we conclude that dH_D is an isometry (with respect to g_D, $g_{H(D)}$ as defined in (15.17)), and that it maps $\mathcal{T}_0(D)$ to $\mathcal{T}_0(H(D))$, $\mathcal{T}_1(D)$ to $\mathcal{T}_1(H(D))$. Thus H is an isometry, and it preserves the splitting of $\mathcal{T}$ into $\mathcal{T}_0 + \mathcal{T}_1$.

Note that the fact that dH_D sends $\mathcal{T}_0(D)$ to $\mathcal{T}_0(H(D))$ is the differentiated version of our earlier result that H sends domains with the same Kobayashi indicatrix to domains with the same indicatrix (by Theorem 13.1).

The preservation of $\mathcal{T}_1$ by dH also has some interesting consequences. Recall that if $I \in \mathcal{I}^\infty$, then $\mathcal{T}_1(I)$ represents the tangent space to $\mathcal{I}^\infty$ at I. If $h \in \mathcal{H}$ and $I \in$ domain H, then H ($\mathcal{I}^\infty \cap$ domain H) is a submanifold of D^∞ which passes through $H(I)$ and has tangent space $\mathcal{T}_1(H(I))$ at $H(I)$. In fact the tangent space at each point of this submanifold will be $\mathcal{T}_1$ at that point. Because H is an isometry, this submanifold will also be totally geodesic, at least if we replace $\mathcal{I}^\infty$ by $\mathcal{I}^\infty_{co}$.

There are plenty of submanifolds that arise this way; there is in fact one going through every point of $\mathcal{D}^\omega$. Indeed, let $D \in \mathcal{R}^\omega$ be given, and let $\rho \in \mathcal{R}^\omega$ be such that $\rho(B_n) = D$. Using Proposition 12.11 (or Theorem 7.5) we get that there is an $h \in \mathcal{H}$ with B_n lying in the domain of H and $h = \hat{\rho}$ on $\widehat{B}_n$. Thus $H(B_n) = D$, and H ($\mathcal{I}^\infty_{co} \cap$ domain H) is a submanifold through D whose tangent space at each of its points is represented by $\mathcal{T}_1$.

These submanifolds provide a foliation of $\mathcal{D}^\omega$. [For this statement the word "foliation" should be given some slack, since all the standard definitions cannot be expected to remain equivalent in infinite dimensions.] That is, if h_1, $h_2 \in \mathcal{H}$, and if H_1 ($\mathcal{I}^\infty \cap$ domain H_1) intersects with H_2 ($\mathcal{I}^\infty \cap$ domain H_2) at $D \in \mathcal{D}^\infty$, then the two submanifolds agree in a neighborhood of D. To prove this let I_1, $I_2 \in \mathcal{I}^\infty$ be such that $H_j(I_j) = D$, $j = 1, 2$, and consider $h_3 = h_2^{-1} \circ h_1$. It is enough to show that h_3 commutes with δ_λ on the connected component of its domain that contains $\widehat{I}_1 \setminus \{(0,0)\}$, so that the restriction of h_3 to this component lies in $\widetilde{\mathcal{H}}$.

Set $\psi = h_3 \left|_{\widehat{I}_1}\right.$, and let ϕ be the map of I_1 onto I_2 such that $\phi \circ \Pi = \Pi \circ \psi$ on $\widehat{I}_1$. A calculation like the one used to prove Proposition 8.1 implies that $\delta\phi(\partial F_{I_2}) = \partial F_{I_1}$. This forces ϕ to be complex homogeneous of degree 1, by an argument like the one used to prove Theorem 3.3. [Alternatively, we could use Theorem 3.4 to reduce to that result.] A small amount of chasing of definitions yields that ψ commutes with δ_λ. Because $\widehat{I}_1 \setminus \{(0,0)\}$ is totally real and has real dimension $2n$, and because h_3 is holomorphic, we conclude that h_3 commutes with δ_λ on the connected component of its domain that contains $\widehat{I}_1 \setminus \{(0,0)\}$, as desired.

Let us now prove Proposition 17.2. The point is to compute $\dot{u}_t$ and $\dot{u}'_t$.

Fix $t \in [0,1]$, $z \in D_t \setminus \{0\}$. From (17.1) we have

$$(17.3)\qquad (\tau'_t(\theta_t(z)))^{\cdot} = dh_{\tau_t(z)}(\dot{\tau}_t(z)),$$

where $\dot{\tau}_t(z)$ is viewed as a tangent vector to $\mathcal{C}$ at $\tau_t(z)$, and similarly the left side is viewed as a tangent vector at $p = \tau'_t(\theta_t(z))$. Evaluate Γ_p on both sides of (17.3); using $\delta h(\Gamma) = \Gamma$, we obtain

$$(17.4)\qquad \Gamma_p((\tau'_t(\theta_t(z)))^{\cdot}) = \Gamma_{\tau_t(z)}(\dot{\tau}_t(z)).$$

Using the definition (6.2) of Γ and also the definition of τ_t we see that the right side vanishes. Similarly the left side is easily computed to give

$$(17.5)\qquad \sum_j (\partial_j F'_t)(\theta_t(z))\dot{\theta}_{t,j}(z) = 0,$$

where $\theta_{t,j}$ denotes the jth component of θ_t. In other words, $(\partial F_t' \mid_{\theta_t(z)})(\dot{\theta}_t(z)) = 0$.

On the other hand we also have $F_t' \circ \theta_t = F_t$ (because $\rho_t, \rho_t' \in \mathcal{R}^\infty$). Differentiating this in t gives

$$\dot{F}_t'(\theta_t(z)) + (dF_t' \mid_{\theta_t(z)})(\dot{\theta}_t(z)) = \dot{F}_t(z).$$

The preceeding calculations imply that the second terms vanishes, and so

$$\dot{F}' \circ \theta_t = F_t, \text{ and hence } \dot{u}_t' \circ \theta_t = \dot{u}_t. \tag{17.6}$$

With this in hand it is easy to prove the proposition. From (15.16) we have that $\dot{u}_t = -Y_t(u_t)$, and so $\dot{u}_t' = -Y_t'(u_t')$ because of (17.6), $u_t' \circ \theta_t = u_t$, and the definition of Y_t'. We know from Proposition 16.10 that $Y_t' \in \mathcal{T}(D_t')$, and also that $Y_t'(u_t')$ characterizes Y_t' among elements of $\mathcal{T}(D_t')$, as noted at the end of Section 15. This implies that Y_t' does represent $\frac{d}{dt} D_t'$, as promised.

18. THE FIRST VARIATION OF THE ENERGY OF A CURVE IN $\mathcal{D}_{co}^{\infty}$

Let D_t, $0 \le t \le 1$, be a smooth curve in $\mathcal{D}_{co}^{\infty}$, and let D_{ts}, $0 \le t \le 1$, $-\epsilon < s < \epsilon$, be a variation of D_t, so that $D_{ts}\,\big|_{s=0} = D_t$.

THEOREM 18.1. *Set $E_s = E(D_{ts})$, with $E(D_{ts})$ given as in (15.4). Then*

$$\frac{d}{ds}E_s = \left[(-1)^n \int\limits_{\partial D_{ts}} 2\dot{u}_{ts}u'_{ts}\, d^c u_{ts} \wedge \left(\frac{1}{2i}\partial\overline{\partial}u_{ts}\right)^{n-1}\right]_{t=0}^{1}$$

(18.2)
$$+\,2(-1)^n \int\limits_0^1 \int\limits_{\partial D_{ts}} u'_{ts}\Big\{-\ddot{u}_{ts} + 2\,\mathrm{sgn}(n-1)\sum_{jk} F_{ts}^{j\overline{k}} c_{ts}\left(\partial\dot{u}_{ts} \wedge \overline{\partial}\dot{u}_{ts}\right)_{j\overline{k}}$$
$$- n\dot{u}_{ts}N_{ts}(\dot{u}_{ts}) + \left(\frac{n+2}{2}\right) N_{ts}(\dot{u}^2)\Big\} d^c u_{ts} \wedge \left(\frac{1}{2i}\partial\overline{\partial}u_{ts}\right)^{n-1} dt.$$

In (18.2) we have employed the following notations. The prime denotes a derivative in s, and the dots denote derivatives in t. As usual u_{ts} is the Green's function on D_{ts}, and $F_{ts} = \exp(2u_{ts})$. We let $F_{ts}^{j\overline{k}}$ denote the matrix-valued function inverse to the complex Hessian of F_{ts}, so that

$$\sum_k F_{ts}^{j\overline{k}}\left(\partial_l\overline{\partial}_k F_{ts}\right) = \delta_l^j.$$

The $\mathrm{sgn}(\cdot)$ is the signum function, which takes the values -1, 0, and $+1$ according to whether its argument is negative, zero, or positive.

Given any differential form α on $\overline{D}_{ts} \setminus \{0\}$ we let $c_{ts}(\alpha)$ denote its "compression" to $T^2(D_{ts})$; i.e., $c_{ts}(\alpha)$ agrees with α on $T^2(D_{ts})$, but $i(V)c_{ts}(\alpha) = 0$ for any section V of $T^1(D_{ts})$. [See Section 15 for the definition of T^1, T^2.] We have written $c_{ts}(\partial\dot{u}_{ts} \wedge \overline{\partial}\dot{u}_{ts})_{j\overline{k}}$ for the $dz_j \wedge d\overline{z}_k$ component of $c_{ts}(\partial\dot{u}_{ts} \wedge \overline{\partial}\dot{u}_{ts})$.

For each t, s, N_{ts} is a linear operator on real-valued functions on ∂D_{ts} defined as follows. Fix t, s, and let $a : \partial D_{ts} \to \mathbf{R}$ be given. Let $f : \Delta \to D_{ts}$ be an extremal mapping into D_{ts}, as discussed in Section 5. Then N_{ts} is defined so that

$$(N_{ts}(a))(f(e^{i\theta})) = \text{the normal derivative at } e^{i\theta} \text{ of the harmonic extension of } a \circ f \text{ to } \Delta.$$

The derivation of (18.2) is somewhat complicated and is omitted.

An immediate consequence of (18.2) is that D_t is a geodesic iff

(18.3)

$$-\ddot{u}_t + 2\,\mathrm{sgn}(n-1)\sum_{jk} F_t^{j\bar{k}} c_t \left(\partial \dot{u}_t \wedge \bar{\partial}\dot{u}_t\right)_{j\bar{k}} - n\dot{u}_t N_t(\dot{u}) + \left(\frac{n+2}{2}\right) N_t(\dot{u}^2) = 0$$

on ∂D_t for $t \in [0,1]$.

A simple example of a geodesic is given as follows. Fix a domain D and a positive constant $a > 0$. Set $D_t = \{z : u(z) < -at\}$ for $t \geq 0$, so that $u_t = u + at$. Clearly (18.3) is satisfied.

Let us consider a couple of special cases in which (18.2) and (18.3) simplify. When $n = 1$ the term with the j, k sum disappears and $N_{ts}(a)$ is just the normal derivative of the harmonic extension of a on D_t. The differential form $(-1)^n d^c u_{ts} \wedge \left(\frac{1}{2i}\partial\bar{\partial}u_{ts}\right)^{n-1}$ in (18.2) reduces to $-d^c u_{ts}$, which is just harmonic measure on ∂D_{ts}. Nonetheless, (18.2) and (18.3) are still pretty complicated.

Suppose now that $n > 1$ but that D_t is a completely circled domain for each t. It is not hard to check that the N-terms in (18.2) and (18.3) drop, and it can be shown that (18.2) reduces to

$$-\ddot{F}_t + \sum_{j,k} F_t^{j\bar{k}} (\partial_j F_t)(\bar{\partial}_k F_t) = 0. \tag{18.4}$$

This can be reexpressed in terms of the Monge-Ampère equation as follows. We define $F_\alpha(z)$ for $\alpha \in \mathbf{C}$, $0 < \mathrm{Re}\,\alpha < 1$, by setting $F_\alpha(z) = F_{\mathrm{Re}\,\alpha}(z)$. Then (18.4) is equivalent to the requirement that $F_\alpha(z)$ satisfy HCMA as a function of α and z (away from $z = 0$). According to Theorem 11.5 this is equivalent to saying that $\alpha \mapsto D_{\mathrm{Re}\,\alpha}$ is a special map of $\{\alpha \in \mathbf{C} : \mathrm{Re}\,\alpha \in (0,1)\}$ into $\mathcal{D}^\infty$, in the sense of Section 11.

You can get short-time existence for the initial-value problem for (18.4) in the real-analytic category using Cauchy-Kovalevski, but it does not appear to be well-posed in larger function spaces. There do not seem to exist any general existence results for the boundary value problem for (18.4) with smooth solutions, but there do exist pretty good weak solutions. It can be shown that they are energy-minimizing under certain fairly reasonable conditions.

Using (18.2) it is not difficult to compute directly that if D_t is a curve in $\mathcal{I}_{co}^\infty$ and if the variation D_{ts} to D_t is orthogonal to $\mathcal{I}_{co}^\infty$ at $s = 0$ (i.e., that $\left(\frac{\partial}{\partial s} D_{ts}\right)\big|_{s=0}$ is represented by an element of $\mathcal{T}_0(D_{to})$), then the variation of the energy (18.2) vanishes. We already knew that this would happen, because of Proposition 16.11.

We can also make a similar calculation for normal variations to a curve in $K^{-1}(I)$ for some fixed $I \in \mathcal{I}_{co}^\infty$. This amounts to computing the second fundamental form for $K^{-1}(I)$, to the extent that this makes sense. (We don't know that $K^{-1}(I)$ is a submanifold of $\mathcal{D}_{co}^\infty$.) It is not clear how to interpret the result, except to notice that it need not vanish and so $K^{-1}(I)$ is not totally geodesic. When $n = 1$ the second fundamental form does have a pretty simple formula, and it implies in particular that a geodesic is contained in the $n = 1$ version of $K^{-1}(I)$ only if it is constant.

In fact, when $n = 1$ K satisfies a certain strict concavity property, but it is not clear if there is any analogue of that when $n > 1$. Part of the difficulty with $n > 1$ is that the second and third terms in (18.3) seem to want to go in opposite directions, and neither appears to win.

An intriguing feature of (18.3) is that the occurence of derivatives of $\dot{u}$ is the same as what one generally expects in several complex variables, i.e., two complex-tangential derivatives (in the second term) versus one "normal" derivative (in the third and fourth terms).

Another interesting observation about (18.3) is that the only parts that are non-local (as functions of u_t) are the $N_t(\dot{u}_t)$ and the $N_t(\dot{u}_t^2)$ pieces.

19. GEOMETRY ON $\mathcal{R}^\infty$

The aim of this section is to put a Riemannian structure on $\mathcal{R}^\infty$ that has the following properties: it should be compatible with the metric on $\mathcal{D}^\infty$ and our identification of $\mathcal{D}^\infty$ with $\mathcal{R}^\infty/\mathcal{G}^\infty$ (see Section 9); it should be compatible with our identification of $\mathcal{R}^\infty(\sigma)$ with $\mathcal{D}^\infty(\sigma(B_n))$ for any $\sigma \in \widehat{\mathcal{S}}$ see Section 11; and it should be, in some sense, Hermitian with respect to the "complex structure" $\mathcal{R}^\infty$ has by dint of being a complex variety.

The absence of a manifold structure on $\mathcal{R}^\infty$ produces the usual inconveniences with trying to make sense of a Riemannian metric on $\mathcal{R}^\infty$. It will be helpful to restrict our attention to $\mathcal{R}^\infty_{co}$, the subset of $\mathcal{R}^\infty$ of maps that send B_n to an element of $\mathcal{D}^\infty_{co}$. As before, this restriction does not play a role in the formal computations.

Our first task is to find a good representation for the tangent space to $\mathcal{R}^\infty_{co}$ at one of its elements. This will be very similar to what we did in Section 15 for $\mathcal{D}^\infty_{co}$.

Fix $\rho \in \mathcal{R}^\infty_{co}$, and let $D = \rho(B_n)$. Let$\mathcal{T}_a(D)$ denote the space of continuous vector fields Y on $\overline{D}$ that are smooth away from the origin, that satisfy $Y(0) = 0$ and $Y(z) \in T^1_z(D)$ for $z \neq 0$, and for which $Y(\rho(\lambda z))$ is holomorphic in $\lambda \in \Delta$ for all $z \in \partial B_n$. Here "a" stands for "augmented", since this space is somewhat larger than $\mathcal{T}(D)$, but "analytic" is also relevant. Notice that $\mathcal{T}_a(D)$ really does depend only on D, and not the particular choice of ρ. The following lemma tells us that $\mathcal{T}_a(D)$ represents the tangent space of $\mathcal{R}^\infty_{co}$ at ρ in a reasonable sense. Recall that $P : T(D) \to T^1(D)$ was defined in Section 15.

LEMMA 19.1. *Let ρ_t, $-\epsilon < t < \epsilon$, $\epsilon > 0$, be a smooth curve in $\mathcal{R}^\infty_{co}$, with $\rho_0 = \rho$, and let $V = \dot\rho_0 \circ \rho^{-1}$. Define Y on $\overline{D}$ by $Y(0) = 0$, $Y(z) = P_z(V(z))$ for $z \neq 0$. Then $Y \in \mathcal{T}_a(D)$, and V is uniquely determined by Y, in the sense that if ρ'_t is another such curve, and V', Y' are defined as above, and if $Y = Y'$, then $V = V'$. Conversely, given any $Y \in \mathcal{T}_a(D)$ there is a smooth curve ρ_t in $\mathcal{R}^\infty_{co}$ so that Y is obtained from $\dot\rho_0$ as above.*

The proof of this is not difficult, but we postpone it until later in this section.

Notice that $\mathcal{T}_a(D)$ is a complex vector space. The complex structure on $\mathcal{T}_a(D)$ is compatible with viewing $\mathcal{R}^\infty$ as a complex subvariety of C^∞_a. For example, suppose that $\alpha \mapsto \rho_\alpha$ is a map of $\{\alpha \in \mathbf{C} : |\alpha| < r\}$, $r > 0$, into $\mathcal{R}^\infty_{co}$ that is holomorphic as a map into C^∞_a. We can define the differential of ρ_α in α in the obvious way, and we can view it as a real-linear map of $\mathbf{C}$ into $\mathcal{T}_a(\rho_\alpha(D))$ for each α, using the recipe in the statement of Lemma 19.1. It is clearly complex linear.

Note that holomorphic maps $\alpha \mapsto \rho_\alpha$ can be produced via exponentiation of holomorphic vector fields, as in the discussion at the end of Section 8. This method is more useful when $\rho \in \mathcal{R}^\omega$.

We shall define a metric on $\mathcal{R}^\infty_{co}$ by specifying a bilinear form on $\mathcal{T}_a(D)$ for each $D \in \mathcal{D}^\infty_{co}$. Despite the absence of a manifold structure on $\mathcal{R}^\infty_{co}$, this notion of a Riemannian metric is good enough to work with in practice, and in particular for defining the energy (or the length) of curves.

Fix ρ, D as before. In order to define this bilinear form we need some more notation. Let $N \in \mathcal{T}_a(D)$ denote the vector field on $\overline{D}$ whose value at $\rho(z)$, $z \neq 0$, is given by

$$N(\rho(z)) = \frac{\partial}{\partial r}\bigg|_{r=1} \rho(rz), \tag{19.2}$$

or, equivalently,

$$N(f(\lambda)) = \frac{\partial}{\partial r}\bigg|_{r=1} f(r\lambda)$$

for each extremal mapping $f : \Delta \to D$ and all $\lambda \in \Delta$. Let $\mathcal{T}_{a1}(D)$ denote the space of $Y \in \mathcal{T}_a(D)$ for which there is a complex-valued function η on $\overline{D}\setminus\{0\}$ such that

$$Y = \eta N \quad \text{on } \overline{D}\setminus\{0\}, \text{ and} \tag{19.3}$$

$$\eta(\rho(\lambda z)) = \eta(\rho(z)) \quad \text{for all } z \in \overline{B}_n \setminus \{0\} \text{ and } \lambda \in \Delta \setminus \{0\}. \tag{19.4}$$

It is easy to check that $\mathcal{T}_a(D)$ is the direct sum of $\mathcal{T}_{a1}(D)$ and $\mathcal{T}_0(D)$, and that $\mathcal{T}_1(D) = \mathcal{T}(D) \cap \mathcal{T}_{a1}(D)$ corresponds to real-valued η's.

Given Y_1, $Y_2 \in \mathcal{T}_a(D)$, we define

$$\begin{aligned} G_\rho(Y_1, Y_2) &= (-1)^n \int_{\partial D} Y_1(u) Y_2(u)\, d^c u \wedge \left(\frac{1}{2i}\partial\overline{\partial} u\right)^{n-1} \\ &\quad + (-1)^n \int_{\partial D} (\operatorname{Im}\eta_1)(\operatorname{Im}\eta_2)\, d^c u \wedge \left(\frac{1}{2i}\partial\overline{\partial} u\right)^{n-1}, \end{aligned} \tag{19.5}$$

where u is the Green's function for D, $Y_j(u)$ denotes the action of Y_j as a vector field on u (as in (3.6)), $j = 1, 2$, and where η_1, η_2 are complex-valued functions on $\overline{D}\setminus\{0\}$ that satisfy (19.4) and

$$Y_j - \eta_j N \in \mathcal{T}_0(D), \ j = 1, 2.$$

We have written G_ρ instead of G_D, even though the definition of G_ρ depends only on D, to emphasize that we view this as giving a Riemannian metric on $\mathcal{R}^\infty$ at ρ.

LEMMA 19.6. *(a) G_ρ is positive definite on $\mathcal{T}_a(D)$;*
(b) $G_\rho(Y_1, Y_2) = g_D(Y_1, Y_2)$ when Y_1, $Y_2 \in \mathcal{T}(D)$;
(c) the orthogonal complement of $\mathcal{T}(D)$ in $\mathcal{T}_a(D)$ with respect to G_ρ is
$\{Y \in \mathcal{T}_{a1}(D) : \mathrm{Re}\,\eta = 0\} = J(\mathcal{T}_1(D))$;
(d) G_ρ is Hermitian on $\mathcal{T}_a$.

In (c) we have let J denote the standard complex structure on $\mathbf{C}^n$, and we write $J(\mathcal{T}_1(D))$ for $\{J(Y) : Y \in \mathcal{T}_1(D)\}$, where $J(Y)$ denotes the action of J on the vector field Y. Note that $J(\mathcal{T}_1(D))$ is the set of $Y \in \mathcal{T}_a(D)$ such that Y is tangent to ∂D on ∂D. Indeed, if $Y \in \mathcal{T}_a(D)$ and $f : \Delta \to D$ is any extremal mapping, then there is a function $g : \overline{\Delta} \to \mathbf{C}$ that is holomorphic on Δ such that

$$Y(f(\lambda)) = g(\lambda) \frac{\partial}{\partial r}\bigg|_{r=1} f(r\lambda).$$

In order for Y to be tangent to ∂D on ∂D we must have that g is imaginary on $\partial\Delta$. This forces g to be constant, and Y to lie in $J(\mathcal{T}_1(D))$, since f was arbitrary.

Part (a) follows from the definitions, and Part (b) from (15.17). For (c) it is helpful to observe that any smooth real-valued function on ∂D can be realized as $X(u)$ for some $X \in \mathcal{T}(D)$. [For this (15.18) is relevant.] The verification of (d) is not difficult but will be postponed until after the proof of Lemma 19.1.

Note that the Riemannian structure on $\mathcal{R}_{co}^\infty$ defined by G is invariant under the action of $\mathcal{G}^\infty$ on the right. That is easy to verify, directly from the definitions. The action of $\mathcal{H}$ on $\mathcal{R}_{co}^\infty$ is also isometric. To prove this it is enough to do the following. Fix $h \in \mathcal{H}$ and $\rho \in \mathcal{R}_{co}^\infty$ with ρ in the domain of H. Set $\rho' = H(\rho)$, $\theta = \rho' \circ \rho^{-1}$, $D = \rho(B_n)$, and $D' = \rho'(B_n)$. It suffices to show that the differential dH_ρ of H at ρ is represented by the linear map from $\mathcal{T}_a(D)$ to $\mathcal{T}_a(D')$ given by $Y \mapsto d\theta(Y)$, since this linear transformation is clearly an isometry with respect to G_ρ and $G_{\rho'}$.

In other words, suppose that ρ_t, $-\epsilon < t < \epsilon$, is a smooth curve in $\mathcal{R}_{co}^\infty$ with $\rho_0 = \rho$, and set $\rho_t' = H(\rho_t)$, $V = \dot\rho_0 \circ \rho^{-1}$, and $V' = \dot\rho_0' \circ (\rho')^{-1}$. Let $P : T(D) \to T^1(D)$ be as in Section 15, and let P' denote its counterpart on D'. Then it suffices to show that

$$Y' = d\theta(Y), \tag{19.7}$$

where $Y' = P'(V')$, $Y = P(V)$.

To check this we use the calculations from the proof of Proposition 17.2. Set $D_t = \rho_t(B_n)$, $D_t' = \rho_t'(B_n)$, and let u, u' be the Green's functions on D, D'. As always we take $F = \exp(2u)$, and similarly for F'.

Set $\theta_t = \rho_t' \circ \rho_t^{-1}$ and $\theta = \rho' \circ \rho^{-1}$, so that

$$\dot\rho_0'(z) = \dot\theta_0(\rho(z)) + d\theta\bigg|_{\rho(z)} (\dot\rho_0(z)) \tag{19.8}$$

for all $z \in B_n \setminus \{0\}$. Using (17.5) we obtain

$$\partial F'\,\big|_{\rho'(z)}\,(\dot\rho_0'(z)) = \partial F'\,\big|_{\rho'(z)} \left(d\theta\,\big|_{\rho(z)}\,(\dot\rho_0(z))\right).$$

Because $\delta\theta(\partial F') = \partial F$ (since ρ, $\rho' \in \mathcal{R}^\infty$) we have

$$\left(\partial F' \,\big|_{\rho'(z)}\right)(\dot\rho'_0(z)) = \left(\partial F \,\big|_{\rho(z)}\right)(\dot\rho_0(z)).$$

We can rewrite this as

$$\left(\partial u' \,\big|_{\rho'(z)}\right)(\dot\rho'_0(z)) = \left(\partial u \,\big|_{\rho(z)}\right)(\dot\rho_0(z)), \tag{19.9}$$

using $F' \circ \rho' = F \circ \rho$. From (15.13) we now conclude that

$$P'_{\rho'(z)}(\dot\rho'_0(z)) = d\theta \,\big|_{\rho(z)} \left(P_{\rho(z)}(\dot\rho_0(z))\right). \tag{19.10}$$

This uses the fact that θ carries extremal mappings into D to extremal mappings into D'. From here (19.7) follows immediately.

Although G is, in some sense, a Hermitian metric on $\mathcal{R}^\infty_{co}$, it cannot be Kähler in any reasonable sense. If it were, it would induce a Kähler metric on $GL(n, \mathbf{C})$ that is left invariant, and it is straightforward to show that there is no such metric on $GL(n, \mathbf{C})$.

Let us now discuss the compatibility of this metric on $\mathcal{R}^\infty_{co}$ with our metric on $\mathcal{D}^\infty_{co}$, first in terms of the identification of $\mathcal{D}^\infty$ with $\mathcal{R}^\infty/\mathcal{G}^\infty$.

This identification is effected by the map $R : \mathcal{R}^\infty \to \mathcal{D}^\infty$. Let us compute the differential of R in terms of the representations for the tangent spaces to $\mathcal{R}^\infty_{co}$, $\mathcal{D}^\infty_{co}$ that we've given. Fix $\rho \in \mathcal{R}^\infty_{co}$ and set $D = \rho(B_n)$. Then dR_ρ is represented by the projection of $\mathcal{T}_a(D)$ onto $\mathcal{T}(D)$ whose kernel is $J(\mathcal{T}_1(D))$. This can be formulated more precisely as follows. Suppose that ρ_t, $-\epsilon < t < \epsilon$, is a smooth curve in $\mathcal{R}^\infty_{co}$ such that $\rho_0 = \rho$, and let $Y \in \mathcal{T}_a(D)$ be associated to $\dot\rho_0$ as in Lemma 19.1. Split Y into $Y' + Y''$, where $Y' \in \mathcal{T}(D)$, $Y'' \in J(\mathcal{T}_1(D))$. Then Y' represents $\frac{d}{dt}\big|_{t=0} D_t$, where $D_t = \rho_t(B_n)$, in the sense of Section 15. This is easily obtained from chasing definitions, using the fact that Y'' is tangent to ∂D on ∂D.

It follows from this description of dR_ρ and Lemma 19.6 that $dR_\rho : \mathcal{T}_a(D) \to \mathcal{T}(D)$ maps the orthogonal complement of its kernel isometrically onto its image, with respect to the inner products G_ρ, g_D. This means that $R : \mathcal{R}^\infty_{co} \to \mathcal{D}^\infty_{co}$ is a Riemannian submersion, which is exactly the right kind of compatibility between the metrics on $\mathcal{R}^\infty_{co}$ and $\mathcal{D}^\infty_{co}$ in connection with the identification of the latter as $\mathcal{R}^\infty_{co}/\mathcal{G}^\infty$.

Consider now the correspondence between the metric induced on $\mathcal{R}^\infty_{co}(\sigma)$ by G and the metric induced on $\mathcal{D}^\infty_{co}(I)$ induced by g, where $\sigma \in \mathcal{S}^\infty$, $I \in \mathcal{I}^\infty_{co}$, and $\sigma(B_n) = I$. This is not really so well defined, because we not only do not have manifold structures on $\mathcal{R}^\infty_{co}(\sigma)$ and $\mathcal{D}^\infty_{co}(I)$, but it is not even so clear how to produce plenty of smooth curves inside these spaces as we can for $\mathcal{R}^\infty_{co}$. (We could do better in the real-analytic category, by exponentiating holomorphic vector fields on $\mathcal{C}$, as in Section 8. We can also do better when $n = 2$, using [L3].) It is however a simple matter of chasing definitions to see that the two metrics correspond at least formally under our identification of $\mathcal{R}^\infty_{co}(\sigma)$ and $\mathcal{D}^\infty_{co}(I)$, because of the way we've set things up.

Let us now prove Lemma 19.1. Let u_t denote the Green's function on $D_t = \rho_t(B_n)$, so that $u_0 = u$. [Once again we abandon temporarily our notation where u_0 denotes $\log|z|$, the Green's function on B_n.]

It is easy to see that $Y \in \mathcal{T}_a(D)$ under the hypotheses of the lemma, using Lemma 15.12. Let us show that V is uniquely determined by Y.

We begin by writing down the conditions that V must satisfy to arise in this manner. Because $\rho_t \in \mathcal{R}^\infty$ we have that

$$u_t \circ (\rho_t \circ \rho^{-1}) = u, \quad \delta(\rho_t \circ \rho^{-1})(\partial u_t) = \partial u, \quad \text{and} \quad \delta(\rho_t \circ \rho^{-1})(\partial\overline{\partial} u_t) = \partial\overline{\partial} u.$$

Differentiating these identities in t and evaluating at $t = 0$ gives

$$V(u) = -\dot{u}_0, \quad L_V(\partial u) = -\partial \dot{u}_0, \quad \text{and} \quad L_V(\partial\overline{\partial} u) = -\partial\overline{\partial}\dot{u}_0$$

on $D \setminus \{0\}$. Substituting the first equation into the other two we get that

$$L_V(\partial u) = \partial(V(u)), \quad L_V(\partial\overline{\partial} u) = \partial\overline{\partial}(V(u)). \tag{19.11}$$

To show that V is uniquely determined by Y, in the sense described in the statement of Lemma 19.1, it suffices to show that if W is a continuous vector field on $\overline{D}$ that is smooth away from the origin, $W(0) = 0$, and if W satisfies (19.11) and $P_z(W(z)) = 0$ for all $z \in D \setminus \{0\}$, then $W \equiv 0$.

Suppose that W has these properties, so that W takes values in T^2 and satisfies $W(u) = 0$ in particular. Set $F = \exp(2u)$, and set $\omega = \frac{1}{2i}\partial\overline{\partial} F$, so that ω is a symplectic form on $D \setminus \{0\}$. Because W satisfies (19.11) and $W(u) = 0$ we get that $L_W(\omega) = 0$, so that W is an ω-symplectic vector field. Therefore there is a function H on $D \setminus \{0\}$ whose symplectic gradient is W — i.e., $dH = i(W)\omega$ — since $D \setminus \{0\}$ is simply connected. [We are assuming that $n > 1$. The lemma is trivial when $n = 1$.]

To finish this argument we need a pair of auxiliary facts. The first is that ω is bounded in a neighborhood of the origin. The second is that $\omega(X_1, X_2) = 0$ if X_1 is a section of T^1 and X_2 is a section of T^2. Both can be verified by using ρ to reduce to their counterparts on B_n.

From the first auxiliary fact we obtain that dH is bounded on a neighborhood of the origin, so that in particular H admits a continuous extension across 0. The second fact implies that dH vanishes on T^1, from which we conclude that $H(\rho(\lambda z))$ is constant in $\lambda \in \Delta \setminus \{0\}$ for all $z \in \partial B_n$. Combining these two observations gives us that H is constant on $\overline{D}$, and hence $W \equiv 0$, as desired.

Next we prove that every $Y \in \mathcal{T}_a(D)$ arises from a curve in $\mathcal{R}^\infty_{co}$ as in Lemma 19.1. We first show that we can reduce to the following.

LEMMA 19.12. *If $Y \in J(\mathcal{T}_1(D))$, then there is a smooth curve $\alpha_t \in \mathcal{G}^\infty$, $-1 < t < 1$, such that α_0 is the identity and $Y = P(V)$, where $V = \dot{\rho}_0 \circ \rho^{-1}$ and $\rho_t = \rho \circ \alpha_t$.*

Let us show why we can reduce to Lemma 19.12. Let $Y \in \mathcal{T}_a(D)$ be arbitrary, and split it into $Y' + Y''$, where $Y' \in \mathcal{T}(D)$ and Y'' lies in $J(\mathcal{T}_1(D))$. Let D_t, $-1 < t < 1$, be a smooth curve in $\mathcal{D}^\infty_{co}$ such that $D_0 = D$ and $\frac{d}{dt}\big|_{t=0} D_t$ is represented by Y' in the sense of Section 15. Let ρ'_t be a smooth curve in $\mathcal{R}^\infty$ such that $\rho'_t(B_n) = D_t$, $\rho_0 = \rho$.

[Such a curve ρ'_t always exists, for the following reason. Set $I_t = K(D_t)$, and $\sigma = S(\rho)$, and let $\Psi_t : I_t \to D_t$ be as in Section 5. There is a smooth curve σ_t in $\mathcal{S}^\infty$ such that $\sigma_0 = \sigma$ and $\sigma_t(B_n) = I_t$; σ_t can be produced using the same argument as in the proof of Theorem 3.4. Take $\rho'_t = \Psi_t \circ \sigma_t$.]

Let $V_1 = \dot\rho'_0 \circ \rho^{-1}$. Then $V_1 - Y'$ must be tangent to ∂D on ∂D, and hence $P(V_1) - Y'$ must be tangent to ∂D. It must also lie in $J(\mathcal{T}_1(D))$, since it is an element of $\mathcal{T}(D)$. Thus we can apply Lemma 9.12 to

$$Y' - P(V_1) + Y'' = Y - P(V_1).$$

If α_t is as in that lemma, then $\rho_t = \rho'_t \circ \alpha_t$ is a smooth curve in $\mathcal{R}^\infty_{co}$ that satisfies $\rho_0 = \rho$ and $P(V) = Y$, $V = \dot\rho_0 \circ \rho^{-1}$.

It remains to prove Lemma 19.12. It is easy to reduce $\rho = \mathrm{id}$, $D = B_n$, and we do.

We are going to obtain α_t by flowing along a vector field W on B_n. That is, we are going to find a vector field W with certain properties, and then take α_t to be the solution of $\dot\alpha_t = W \circ \alpha_t$ and $\alpha_0 = \mathrm{id}$. In order that α_t have the features required by Lemma 19.12 we need to find W which is tangent to ∂B_n on ∂B_n, W is complex homogeneous of degree 1, W is Hamiltonian with respect to $\omega_0 = \frac{1}{2i}\partial\overline{\partial}F_0$, $F_0(z) = \Sigma|z_j|^2$, and $P(W) = Y$, where $Y \in J(\mathcal{T}_1(B_n))$ is given, and $P : T(B_n) \to T^1(B_n)$ is as in Section 15.

Let $Y_k(z)$, $k = 1, 2, \dots, n$, denote the complex components of Y. Because $Y \in J(\mathcal{T}_1(D))$ there is a function η on $B_n \setminus \{0\}$ that takes values in imaginary numbers and satisfies $\eta(\lambda z) = \eta(z)$ for all $\lambda \in \Delta \setminus \{0\}$ and $Y_k(z) = \eta(z) z_k$ for $z \in B_n \setminus \{0\}$.

We take W to be the ω_0-symplectic gradient on $B_n \setminus \{0\}$ of

$$H(z) = \frac{1}{2i}\,\eta(z)|z|^2. \tag{19.13}$$

In other words, W is the vector field such that

$$dH(V) = \omega_0(W, V)$$

for all other vector fields V.

Let W_k, V_k denote the complex components of W, V. We can rewrite the preceeding equation as

$$2\,\mathrm{Re}\left(\sum_k V_j \partial_j H\right) = \frac{1}{2i}\sum_k \left(W_j\overline{V}_j - V_j\overline{W}_j\right) = -\,\mathrm{Im}\left(\sum_j V_j\overline{W}_j\right).$$

From here we get that $W_j = 2i\overline{\partial}_j H$, and hence

$$W_j(z) = (\overline{\partial}_j\eta)(z)|z|^2 + \eta(z) z_j. \tag{19.14}$$

Let's check that W has the required properties. It is certainly Hamiltonian with respect to ω_0, and it is easy to check that it is complex homogeneous of degree 1. Observe that the first term on the right in (19.14) lies in S_z^2 because of the homogeneity identity

$$\sum \overline{z}_j \cdot \overline{\partial}_j \eta(z) = \frac{\partial}{\partial \overline{\lambda}} \Big|_{\lambda=1} \eta(\lambda z) = 0.$$

This implies that $P(W) = Y$, and in particular that W is tangent to ∂B_n on ∂B_n.

Thus we can obtain α_t by solving $\dot{\alpha}_t = W \circ \alpha_t$, $\alpha_0 = \mathrm{id}$. We can do this for all $t \in \mathbf{R}$; there are no blow-ups, because W is tangent to $\{z \in B_n : |z| = r\}$, $0 < r < 1$, on this set, and so α_t maps this set to itself for all t.

This finishes the proof of Lemmas 19.12 and 19.1. It remains to check (d) of Lemma 19.6, which we do by exhibiting a suitable formula for $G_\rho(Y, Y')$.

Let Y, Y' be two elements of $\mathcal{T}(D)$. Let η, η' be chosen so that they satisfy (19.4) and so that $Y_0 = Y - \eta N$, $Y_0' = Y' - \eta' N$ lie in $\mathcal{T}_0(D)$. Then we have that

$$\begin{aligned} G_\rho(Y, Y') &= (-1)^n \int_{\partial D} Y_0(u) Y_0'(u) d^c u \wedge \left(\frac{1}{2i}\, \partial\overline{\partial} u\right)^{n-1} \\ &\quad + (-1)^n \int_{\partial D} \mathrm{Re}(\eta \overline{\eta'}) d^c u \wedge \left(\frac{1}{2i}\, \partial\overline{\partial} u\right)^{n-1}. \end{aligned} \tag{19.15}$$

This follows from (19.5) using the orthogonality of $\mathcal{T}_0(D)$ and $\mathcal{T}_1(D)$ (Proposition 16.9) and the fact that $N(u) = 1$ on ∂D, which can be derived from the definition of N. The second term is obviously Hermitian, since η depends complex-linearly on Y, but we must compute the first term some more.

There are continuous complex-valued functions g, g' on D' such that $Y_0 = gN$, $Y_0' = g'N$, $g(0) = 0 = g'(0)$, and $g(\rho(\lambda z))$ and $g'(\rho(\lambda z))$ are holomorphic in $\lambda \in \Delta$ for $z \in B_n$. When we write gN we are using the natural complex structure on vector fields on $\mathbf{C}^n$, but we still view it as a "real" vector field, i.e., as defining a directional derivative as in (3.6).

Let g_1 and g_2 denote the real and imaginary parts of g, and similarly for g_1', g_2'. Then $Y_0(u) = g_1 N(u)$ and $Y_0'(u) = g_1' N(u)$ on ∂D because $Y_0 - g_1 N$ and $Y_0' - g_1' N$ are tangent to ∂D on ∂D. Hence the first term on the right side of (19.15) equals

$$(-1)^n \int_{\partial D} g_1 g_1' d^c u \wedge \left(\frac{1}{2i}\, \partial\overline{\partial} u\right)^{n-1}, \tag{19.16}$$

since $N(u) = 1$ on ∂D.

This is also equal to

$$\frac{1}{2}\,(-1)^n \int_{\partial D} \mathrm{Re}(g\overline{g'}) d^c u \wedge \left(\frac{1}{2i}\, \partial\overline{\partial} u\right)^{n-1}. \tag{19.17}$$

To see this we use ρ to make a change of variables in (19.16), and we get that (19.16) equals

$$(-1)^n \int_{\partial B_n} (g_1 \circ \rho)(g_1' \circ \rho) d^c u_0 \wedge \left(\frac{1}{2i}\, \partial\bar{\partial} u_0\right)^{n-1},$$

where u_0 is once again the Green's function on B_n. Using rotation invariance and Fubini's theorem we can write this as

$$(-1)^n \int_{\partial B_n} \frac{1}{2\pi} \int_0^{2\pi} g_1(\rho(e^{i\theta} z)) g_1'(\rho(e^{i\theta} z)) d\theta \; d^c u_0 \wedge \left(\frac{1}{2i}\, \partial\bar{\partial} u_0\right)^{n-1}, \tag{19.18}$$

where z denotes the integration variable for the outer integral. Because $g(\rho(\lambda z))$ and $g'(\rho(\lambda z))$ are holomorphic in $\lambda \in \Delta$ and vanish at $\lambda = 0$ we have that

$$\frac{1}{2\pi} \int_0^{2\pi} g_1(\rho(e^{i\theta} z)) g_1'(\rho(e^{i\theta} z)) d\theta = \frac{1}{2\pi} \int_0^{2\pi} \frac{1}{2}\, \mathrm{Re}(g(\rho(e^{i\theta} z)) \overline{g'(\rho(e^{i\theta} z))}) d\theta.$$

If we substitute this back into (19.18), use Fubini to get rid of the θ-integral, and then undo the change of variables, we get (19.17).

Thus we can replace the first term on the right in (19.15) by (19.17). The resulting formula makes it clear that G_ρ is Hermitian, as desired.

There are other ways to write this formula, which we now record for future use. Because $du(N) = 1$ on ∂D and $du(JN) = 0$, we have $\partial u(N) = \frac{1}{2}$ and hence

$$\begin{aligned} G_\rho(Y, Y') = {} & 2(-1)^n \,\mathrm{Re} \int_{\partial D} \partial u(Y_0) \bar{\partial} u(Y_0') d^c u \wedge \left(\frac{1}{2i}\, \partial\bar{\partial} u\right)^{n-1} \\ & + 4(-1)^n \,\mathrm{Re} \int_{\partial D} \partial u(\eta N) \bar{\partial} u(\eta' N) d^c u \wedge \left(\frac{1}{2i}\, \partial\bar{\partial} u\right)^{n-1}. \end{aligned} \tag{19.19}$$

Let's rewrite this in terms of Y, Y', without Y_0, Y_0', η, η'. To do this we use the fact that

$$\begin{aligned} \frac{1}{2\pi} \int_0^{2\pi} \partial u(Y)(\rho(e^{i\theta} z)) d\theta &= \frac{1}{2\pi} \int_0^{2\pi} \frac{1}{2}\, (g(\rho(e^{i\theta} z)) + \eta(\rho(e^{i\theta} z))) d\theta \\ &= \frac{1}{2}\, \eta(\rho(z)) = (\partial u(\eta N))(\rho(z)), \end{aligned}$$

which follows from the holomorphicity of $g(\rho(\lambda z))$ in λ and $g(0) = 0$. This and an argument like the one employed to obtain the equality of (19.16) and (19.17) gives

$$\begin{aligned} & \int_{\partial D} \partial u(Y) \bar{\partial} u(Y') d^c u \wedge \left(\frac{1}{2i}\, \partial\bar{\partial} u\right)^{n-1} \\ & = \int_{\partial D} (\partial u(Y_0) \bar{\partial} u(Y_0) + \partial u(\eta N) \bar{\partial} u(\eta' N)) \; d^c u \wedge \left(\frac{1}{2i}\, \partial\bar{\partial} u\right)^{n-1} \end{aligned}$$

(i.e., the cross terms vanish), and hence

(19.20)

$$G_\rho(Y,Y') = 2(-1)^n \operatorname{Re} \int_{\partial D} \partial u(Y)\overline{\partial} u(Y') d^c u \wedge \left(\frac{1}{2i}\,\partial\overline{\partial} u\right)^{n-1} + 2(-1)^n \times$$

$$\operatorname{Re} \int_{\partial D} \left(\frac{1}{2\pi}\int_0^{2\pi} r_{D,\theta}(\partial u(Y))d\theta\right)\left(\frac{1}{2\pi}\int_0^{2\pi} r_{D,\phi}(\overline{\partial} u(Y'))d\phi\right) d^c u \wedge \left(\frac{1}{2i}\,\partial\overline{\partial} u\right)^{n-1},$$

where $r_{D,\theta}$ is the map of $\overline{D}$ onto itself such that $r_{D,\theta}(\rho(z)) = \rho(e^{i\theta}z)$.

20. ANOTHER APPROACH TO RIEMANNIAN GEOMETRY ON $\mathcal{R}^\infty$

So far we have dealt with the problem of putting Riemannian metrics on $\mathcal{D}^\infty$, $\mathcal{R}^\infty$, despite the absence of a manifold structure, by restricting ourselves to strongly convex domains and using the work of Lempert to provide adequate control of the analysis. Now we take a different tack, which is closer to the idea of viewing $\mathcal{R}^\infty$ as a complex variety in $\mathcal{C}_a^\infty$: we look for a smooth metric on $\mathcal{C}_a^\infty$ which restricts to the metric we want on $\mathcal{R}^\infty$, at least formally. Actually, we are going to work with $\widehat{\mathcal{R}}^\infty$ instead of $\mathcal{R}^\infty$, and this will give us nicer formulae. For simplicity we'll take the ambient space to be $C^\infty(\partial\widehat{B}_n, \mathcal{C})$ instead of $\widehat{\mathcal{C}}_a^\infty$. It turns out that the natural metric to put on $C^\infty(\partial\widehat{B}_n, \mathcal{C})$ is nonnegative but not positive definite, although it induces a positive-definite metric on $\widehat{\mathcal{R}}^\infty$.

We identify the tangent bundle of $C^\infty(\partial\widehat{B}_n, \mathcal{C})$ with $C^\infty(\partial\widehat{B}_n, \mathcal{C}) \times C^\infty(\partial\widehat{B}_n, \mathcal{C})$ in the obvious way. Fix $\psi \in C^\infty(\partial\widehat{B}_n, \mathcal{C})$ so that the tangent space to $C^\infty(\partial\widehat{B}_n, \mathcal{C})$ at ψ is identified with $C^\infty(\partial\widehat{B}_n, \mathcal{C})$. Let α, $\alpha' \in C^\infty(\partial\widehat{B}_n, \mathcal{C})$ be any two such tangent vectors.

Using our identification of $\mathcal{C}$ with $\mathbf{C}^n \times \mathbf{C}^n$ we can split ψ into two sets of components ψ_1, ψ_2, each of which can be further subdivided into n components $\psi_{1,j}$, $\psi_{2,j}$, $j = 1, \ldots, n$. We can do the same to α, α' to get α_1, α_2, etc.

We associate to α the function $f_{\psi,\alpha} : \partial\widehat{B}_n \to \mathbf{C}$ defined by

$$f_{\psi,\alpha} = \sum \psi_{2,j}\alpha_{1,j}, \tag{20.1}$$

and similarly for $f_{\psi,\alpha'}$. The reason for this will become clearer later.

Set $\mu_0 = \delta i_{\widehat{B}_n}(\mu)$, $\Gamma_0 = \delta i_{\widehat{B}_n}(\Gamma)$, where μ, Γ are as in Section 6. We define our semidefinite metric on $C^\infty(\partial B_n, \mathcal{C})$ by

(20.2)

$$\widehat{G}_\psi(\alpha, \alpha') = 4(-1)^n \operatorname{Re} \int_{\partial\widehat{B}_n} f_{\psi,\alpha}\overline{f_{\psi,\alpha'}}(\operatorname{Im}\Gamma_0) \wedge \mu_0^{n-1} + 4(-1)^n \times$$

$$\operatorname{Re} \int_{\partial\widehat{B}_n} \left(\frac{1}{2\pi}\int_0^{2\pi} (f_{\psi,\alpha} \circ \delta_{e^{i\theta}})\, d\theta\right) \left(\frac{1}{2\pi}\int_0^{2\pi} \overline{(f_{\psi,\alpha'} \circ \delta_{e^{i\phi}})}\, d\phi\right) (\operatorname{Im}\Gamma_0) \wedge \mu_0^{n-1},$$

where δ_λ is as in Section 11. The similarity between this and (19.20) is intentional.

This metric has the nice properties of being Hermitian and nonnegative, even if it is degenerate. To check nonnegativity we make a change of variables. Suppose that k is any function on $\partial\widehat{B}_n$, and define $\tau_0 : \mathbf{C}^n \to \mathcal{C}$ by $\tau_0(z) = (z, \overline{z})$. Then

$$(20.3)\quad (-1)^n \int_{\partial\widehat{B}_n} k(\operatorname{Im}\Gamma_0) \wedge \mu_0^{n-1} = \frac{1}{2}(-1)^n \int_{\partial B_n} k \circ \tau_0 \; d^c u_0 \wedge \left(\frac{1}{2i}\partial\overline{\partial}u_0\right)^{n-1},$$

where u_0 is the Green's function for B_n. To see this we use (6.5) to get that $\delta\tau_0(\Gamma) = \partial F_0$, which equals ∂u_0 on ∂B_n, so that $\delta\tau_0(\operatorname{Im}\Gamma_0) = \frac{1}{2}d^c u_0$, and we also use (6.7) to obtain $\delta\tau_0(\mu) = \frac{1}{2i}\partial\overline{\partial}F_0$, which pulls back to $\frac{1}{2i}\partial\overline{\partial}u_0$ on ∂B_n. It is not difficult to check that the right side is nonnegatiave if k is.

Another nice property of $\widehat{G}_\psi$ is that it is a homogeneous quadratic polynomial as a function of ψ.

Let us check that this metric is compatible with the one discussed in the preceeding section. In other words, we want to check that the mapping $\rho \mapsto \hat{\rho}$ of $\mathcal{R}^\infty$ to $\widehat{\mathcal{R}}^\infty \subseteq C^\infty(\widehat{B}_n, \mathcal{C})$ defines an isometry. We interpret this to mean the following. Fix $\rho \in \mathcal{R}^\infty$, and let ρ_t, $-\epsilon < t < \epsilon$, be a smooth curve in $\mathcal{R}^\infty$ with $\rho_0 = \rho$. If we assign a length to $\dot{\rho}_0$ in accordance with the preceeding section, and also to $\frac{d}{dt}\big|_{t=0} \hat{\rho}_t$ using $\widehat{G}$, then these two lengths are the same.

This is not hard to verify. Let V denote the vector field on $\overline{D}$ given by $V = \dot{\rho}_0 \circ \rho^{-1}$, and let Y be the vector field on $\overline{D}$ given by $Y = P(V)$, where $P : T(D) \to T^1(D)$ is as in Section 15. Then the length of $\dot{\rho}_0$ is given by $(G_\rho(Y,Y))^{\frac{1}{2}}$, where G_ρ is given by (19.20). We may replace Y by V in the right side of (19.20), because $\partial u(Y) = \partial u(V)$ (since $V - Y$ is a section of $T^2(D)$).

On the other hand, when we compute the length of $\frac{d}{dt}\big|_{t=0} \hat{\rho}_t$ using $\widehat{G}$, we should take $\psi = \hat{\rho}$ and $\alpha = \alpha' = \frac{d}{dt}\big|_{t=0} \hat{\rho}_t$, and we get from the definition (7.1) of $\hat{\rho}$ that $\psi_{2,j} \circ \tau_0 = (\partial_j F) \circ \rho$ and that $\alpha_1 \circ \tau_0 = V \circ \rho$ where $F = e^{2u}$ and u is the Green's function on D, as always. Thus $f_{\psi,\alpha} \circ \tau_0 = (\partial F(V)) \circ \rho$, which equals $(\partial u(V)) \circ \rho$ on ∂B_n.

The equality between (19.20) and (20.2) with these choices is a consequence of two changes of variables and a little calculation. The first change of variables is that if k is a function on ∂D, then

$$\int_{\partial D} k d^c u \wedge \left(\frac{1}{2i}\partial\overline{\partial}u\right)^{n-1} = \int_{\partial B_n} k \circ \rho \; d^c u_0 \wedge \left(\frac{1}{2i}\partial\overline{\partial}u_0\right)^{n-1}.$$

This follows from $\delta\rho(\partial u) = \partial u_0$ and $\delta\rho(\partial\overline{\partial}u) = \partial\overline{\partial}u_0$, which hold since $\rho \in \mathcal{R}^\infty$. The second change of variables is the one in (20.3).

Next we check that the natural actions of $\mathcal{H}$ and $\widehat{\mathcal{G}}^\infty$ on $C^\infty(\partial\widehat{B}_n, \mathcal{C})$ by composition on the left and right preserve $\widehat{G}$. This extends an observation of the preceeding section, which dealt with the action of $\mathcal{H}$ and $\mathcal{G}^\infty$ on $\mathcal{R}^\infty_{co}$.

The fact that the action of $\widehat{\mathcal{G}}^\infty$ on $C^\infty(\partial\widehat{B}_n, \mathcal{C})$ preserves $\widehat{G}$ is immediate from the definitions, because elements of $\widehat{\mathcal{G}}^\infty$ preserve $\partial\widehat{B}_n$ and Γ_0, μ_0. The corresponding fact for h is a little more subtle, because we have to be more careful about what happens to $f_{\psi,\alpha}$.

Fix $h : \Omega \mapsto \mathcal{C}$ in $\mathcal{H}$, and recall that $\widehat{H}$ is defined on the $\psi \in C^\infty(\partial\widehat{B}_n, \mathcal{C})$ with $\psi(\partial\widehat{B}_n) \subseteq \Omega$ by $\widehat{H}(\psi) = h \circ \psi$. Given such a ψ, if we view $\alpha \in C^\infty(\partial\widehat{B}_n, \mathcal{C})$ as a tangent vector at ψ, then

$$d\widehat{H}_\psi(\alpha) = \frac{d}{dt}\Big|_{t=0} h(\psi + t\alpha)$$

defines an element β of $C^\infty(\partial\widehat{B}_n, \mathcal{C})$, which we view as a tangent vector to $C^\infty(\partial\widehat{B}_n, \mathcal{C})$ at $\phi = \widehat{H}(\psi) = h \circ \psi$. We can rewrite β as

$$\beta(w) = dh_{\psi(w)}(\alpha(w)), \qquad w \in \partial\widehat{B}_n. \tag{20.4}$$

On the right hand side we view $\alpha(w)$ as being a tangent vector to $\mathcal{C}$ at $\psi(w)$.

The key point is that

$$f_{\phi,\beta} = f_{\psi,\alpha}. \tag{20.5}$$

This follows from the condition $\delta h(\Gamma) = \Gamma$.

To understand this better it is helpful to rewrite $f_{\psi,\alpha}$ in terms of Γ. If we think of ψ geometrically, as a mapping, rather than merely as an element of a vector space, then it is natural to think of an infinitesimal variation of ψ as being given by a vector field along its image. Thus if we fix $w \in \partial\widehat{B}_n$, then we should view $v = \alpha(w)$ as an element of the tangent space to $\mathcal{C}$ at $\psi(w)$, in which case we can write

$$f_{\psi,\alpha}\Big|_w = \left(\Gamma\Big|_{\psi(w)}\right)(v). \tag{20.6}$$

This makes (20.5) more transparent.

Let us come back now and say a little more about how one might arrive at formula (20.2) for $\widehat{G}_\psi$. We introduced it as a metric on $C^\infty(\partial\widehat{B}_n, \mathcal{C})$ that would induce the same metric on $\widehat{\mathcal{R}}^\infty$ as the one we discussed in the preceeding section. It is not hard to arrive at pretty much the same formula from more direct considerations.

In order to decide what a good metric for $\widehat{\mathcal{R}}^\infty$ might be we should first decide what the tangent space should look like. Fix $\psi \in \widehat{\mathcal{R}}^\infty$, and set $N = \psi(\widehat{B}_n) \setminus \{(0,0)\}$. An infinitesimal variation of ψ in $\widehat{C}^\infty_a$ can be represented by a vector field V along N that tends to 0 as you move toward $(0,0)$ and which has certain holomorphicity properties. For V to represent an infinitesimal variation of ψ inside $\widehat{\mathcal{R}}^\infty$, and not just $\widehat{C}^\infty_a$, it is necessary to respect the constraint $\delta\psi(\Gamma) = \delta i_{\widehat{B}_n}(\Gamma)$. The linearized condition is

$$\delta i_N(L_V(\Gamma)) = 0. \tag{20.7}$$

This is somewhat of an abuse of notation, because the Lie derivative L_V is usually defined only when the domain of V is open. The pull-back of $L_V\Gamma$ to N does make sense, however. To make this precise, and also to get a nice formula, let us show that if $\widetilde{V}$ is any vector field on a neighborhood of N that agrees with V on N, then $\delta i_N(L_{\widetilde{V}}(\Gamma))$ depends only on V and not $\widetilde{V}$.

Using the well-known formula

$$L_{\widetilde{V}} = d \circ i(\widetilde{V}) + i(\widetilde{V}) \circ d,$$

where $i(\widetilde{V})$ denotes interior multiplication (see [Wa], p. 70), we can write

$$\delta i_N(L_{\widetilde{V}}(\Gamma)) = d\left(\delta i_N(i(\widetilde{V})\Gamma)\right) + \delta i_N(i(\widetilde{V})d\Gamma).$$

The right side is clearly independent of the choice of extension $\widetilde{V}$, and we define the left side of (20.7) by

$$\delta i_N(L_V(\Gamma)) = d(\delta i_N(i(V)\Gamma)) + \delta i_N(i(V)d\Gamma). \tag{20.8}$$

Of course $\delta i_N(i(V)\Gamma) = \Gamma(V)$ is just a complex-valued function on N.

To analyze (20.7) we split V into $V_1 + V_2$, where V_1 is a section of the tangent bundle TN of N, and where V_2 is a section of $\widehat{J}(TN)$, with $\widehat{J}$ denoting the complex structure on $\mathcal{C}$. Every vector field along N admits such a splitting because N is totally real (since $\psi \in \widehat{\mathcal{R}}^\infty$).

Recall from Section 6 that $d\Gamma = -2i\gamma$, $\gamma = \mu + i\nu$. A necessary condition for (20.7) to hold is that

$$0 = d(\delta i_N(L_V(\Gamma))) = d\{\delta i_N(i(V)(-2i\gamma))\}. \tag{20.9}$$

Because γ is complex linear and N is ν-Lagrangian (since $\psi \in \widehat{\mathcal{R}}^\infty$) we have that

$$\delta i_N(i(V_1)\nu) \equiv 0, \quad \delta i_N(i(V_2)\mu) \equiv 0, \tag{20.10}$$

and so (20.9) is equivalent to

$$d\{\delta i_N(i(V_1)\mu)\} = 0 \quad \text{and} \tag{20.11}$$

$$d\{\delta i_N(i(V_2)\nu)\} = 0. \tag{20.12}$$

Using again the complex linearity of γ we have that (20.12) is the same as

$$d\{\delta i_N(i(\widehat{J}V_2)\mu)\} = 0. \tag{20.13}$$

Set $\mu_N = \delta i_N(\mu)$. This is a symplectic form on N; it is nondegenerate because $\delta\psi(\mu_N) = \delta i_{\widehat{B}_n}(\gamma)$ is. Thus the preceeding calculations show that (20.9) holds if and only if V_1 and $\widehat{J}V_2$ are μ_N-symplectic vector fields on N.

Remember that we are trying to find a natural metric on $\widehat{\mathcal{R}}^\infty$, which is to say a natural way to measure the "length" of V, viewed as an element of the tangent space to $\widehat{\mathcal{R}}^\infty$ at ψ. Since we know that V_1 and $\widehat{J}V_2$ must be μ_N-symplectic, and hence μ_N-Hamiltonian (as long as $n > 1$, but the $n = 1$ case is sort of trivial for this), it is natural to do this by taking an L^2 norm of the appropriate potentials.

Because V is supposed to satisfy (20.7), and not just (20.9), we can get nice formulae for potentials of V_1 and JV_2. From (20.7) and (20.8) we obtain

$$d(\Gamma(V)) = -\delta i_N(i(V)d\Gamma) = 2\sqrt{-1}\delta i_N(i(V)\gamma). \tag{20.14}$$

Using (20.10) and the complex linearity of γ we have

$$\begin{aligned} \delta i_N(i(V)\gamma) &= i(V_1)\mu_N + \delta i_N(i(V_2)\gamma) \\ &= i(V_1)\mu_N - \sqrt{-1}\delta i_N(i(\widehat{J}V_2)\gamma) \\ &= i(V_1)\mu_N - \sqrt{-1}i(\widehat{J}V_2)\mu_N. \end{aligned} \tag{20.15}$$

Combining (20.14) and (20.15), we see that $\widehat{J}V_2$ and V_1 are the μ_N-symplectic gradients of the real and imaginary parts of $\Gamma(V)$, except for some nonzero multiplicative constants.

Thus it is natural to define the length of V in terms of an L^2 norm of $\Gamma(V)$. We may as well make it an L^2 norm obtained by integrating on ∂N; this is not unreasonable because V is determined by its values on ∂N (since $V \circ \psi$ lies in C_a^∞). We have to decide on which measure to integrate against, but in order to have nice invariance properties we should define it in terms of Γ and γ. An obvious choice is to take

$$(\text{length of } V)^2 = (-1)^n \int_{\partial N} |\Gamma(V)|^2 \operatorname{Im}(\Gamma_N) \wedge \mu_N^{n-1}, \tag{20.16}$$

where $\Gamma_N = \delta i_N(\Gamma)$, and the $(-1)^n$ is there to make the integral positive.

This it is pretty much the same as (20.2). We can use ψ as a change of variables in (20.16) to write it as in integral on $\partial\widehat{B}_n$, to wit

$$(-1)^n \int_{\partial\widehat{B}_n} (|\Gamma(V)|^2) \circ \psi \operatorname{Im}(\Gamma_0) \wedge \mu_0^{n-1}. \tag{20.17}$$

If we take $\alpha = \alpha' = V \circ \psi$, then $\Gamma(V)\,|_{\psi(w)} = f_{\psi,\alpha}\,|_w$ for all $w \in \partial\widehat{B}_n$, as observed in (20.6). With these choices (20.17) is the same as (20.2) except for a factor of 4 and the second term in (20.2). These adjustments came from the desire to have agreement with the metric on $\mathcal{R}^\infty$ discussed in Section 19.

The preceeding derivation of (20.16) is a variation of a similar discussion in Section 7 in [S].

21. A FEW REMARKS ABOUT THE HERMITIAN GEOMETRY ON $\widehat{\mathcal{R}}^\infty$

The main observation of this section is that $\widehat{\mathcal{R}}^\infty$ has non-positive bisectional curvature, at least if we permit ourselves to interpret this statement with some liberality.

Throughout this section "curvature" will refer to curvature from Hermitian geometry rather than Riemannian geometry. The reader may find it helpful to consult with Chapter 3 of [Wl] or p. 37–39 of [Ko].

Let us recall a few facts about Hermitian geometry. Let M be a complex manifold, and let E be a holomorphic vector bundle over M. Given any Hermitian metric on E there is a unique connection on E that is compatible with both the metric and the complex structure; see [Wl] for both the precise statement and its proof.

Let $\langle \cdot, \cdot \rangle$ be a Hermitian metric on E, and let Θ denote the curvature of the associated connection. Then Θ can be viewed as a $(1,1)$-form that takes values in the bundle of endomorphisms of E. We say that the curvature is positive, negative, etc. if, for each section e of E,

$$\langle \Theta e, e \rangle$$

is positive, negative, etc. as a $(1,1)$-form on M.

Let E' be a holomorphic subbundle on E, so that $\langle \cdot, \cdot \rangle$ provides a Hermitian metric on E' too. If Θ' denotes the curvature of the Hermitian connection on E' (which can be obtained from the one on E by orthogonal projection), then for any section f of E'

$$\langle \Theta f, f \rangle - \langle \Theta' f, f \rangle \tag{21.1}$$

is a nonnegative $(1,1)$-form on M. See, e.g., [Ko], p. 39.

Another (closely related) fact is that E has nonpositive curvature if and only if $\langle e, e \rangle$ is plurisubharmonic for any local holomorphic section e of E. This is basically well known, and it can be verified by direct computation using local co-ordinates and frames.

An important special case occurs when we take E to be the tangent bundle for M. A complex manifold equipped with a Hermitian metric on its tangent bundle is called a Hermitian manifold. It is said to have bisectional curvature that is positive, negative, etc. when the curvature of the tangent bundle is positive, negative, etc., in the sense discussed above. Observe that nonnegativity of (21.1) implies that a complex submanifold of a Hermitian manifold has nonpositive curvature if the original manifold does.

Let us apply these considerations to the situation of the preceeding section. We first look at

$$E = C^\infty(\partial\widehat{B}_n, \mathcal{C}) \times C^\infty(\partial\widehat{B}_n, \mathcal{C})$$

as a trivial vector bundle over $M = C^\infty(\partial\widehat{B}_n, \mathcal{C})$, equipped with the metric $\widehat{G}$ as in (20.2). Because $\widehat{G}$ is semidefinite the usual definitions from Hermitian differential geometry do not apply, but we can still decide to say that E has non-positive curvature if holomorphic sections always have plurisubharmonic norm. This turns out to be true.

Let us be more precise. Let $\psi(w)$, $\alpha(w)$ be maps from some open subset of $\mathbf{C}^m$ (for some m) into $C^\infty(\partial\widehat{B}_n, \mathcal{C})$. The w here denotes the parameter in $\mathbf{C}^m$, so that $\psi(w)$ is, for each w, a $\mathcal{C}$-valued function on $\partial\widehat{B}_n$. We view $\psi(w)$ as defining a map into M and $\alpha(w)$ as being a section of E along $\psi(w)$. If $\psi(w)$ and $\alpha(w)$ are both holomorphic in w, then $\widehat{G}_{\psi(w)}(\alpha(w), \alpha(w))$ is plurisubharmonic in w. This follows easily from (20.2) and the fact that the absolute value of a holomorphic function is plurisubharmonic.

In particular this is true when $\psi(w)$ takes values in $\widehat{\mathcal{R}}^\infty$ and $\alpha(w)$ lives in the (formal) tangent space to $\widehat{\mathcal{R}}^\infty$ at $\psi(w)$. We can again interpret this to mean that $\widehat{\mathcal{R}}^\infty$ has nonpositive curvature, although we are on somewhat shaky ground here, since we do not have a manifold structure for $\widehat{\mathcal{R}}^\infty$. However, we do at least have that $\widehat{G}$ induces a definite metric on the formal tangent bundle of $\widehat{\mathcal{R}}^\infty$.

We can make the ground a little less shaky for $\widehat{\mathcal{R}}^\omega$ using observations of Section 8. In particular we can generate plenty of holomorphic maps in $\widehat{\mathcal{R}}^\omega$ by exponentiation. Let $\mathcal{V}_0$ denote the vector space of functions on $\overline{B}_n$ defined by (8.11). We can view $\mathcal{V}_0$ as a closed subspace of the space of complex-valued real-analytic functions on ∂B_n, from which it inherits a natural topology. There is a kind of exponential mapping from an open subset of $\mathcal{V}_0$ containing the origin into $\widehat{\mathcal{R}}^\omega$, which is defined as follows.

Fix $H \in \mathcal{V}_0$. From Section 8 we know that there is an $\epsilon > 0$ and a holomorphic function $G : \widehat{B}_{n,2\epsilon} \to \mathbf{C}$, $\widehat{B}_{n,2\epsilon}$ as in (7.4), that satisfies (8.6), (8.8), $G(0,0) = 0$, and $H(z) = G(z, \overline{z})$. We saw in Section 8 that this implies that the γ-symplectic gradient W of G satisfies $L_W\Gamma = 0$, and that this implies the existence of $t_0 > 0$ and a 1-parameter family $g_t : \widehat{B}_{n,\epsilon} \to \mathcal{C}$, $-t_0 < t < t_0$, of maps such that $g = \mathrm{id}$, $\dot{g}_t = W \circ g_t$, and $\delta g_t(\gamma) = \gamma$, $\delta g_t(\Gamma) = \Gamma$ for each t. It is easy to see that $t_0 > 1$ for an open subset of H's in $\mathcal{V}_0$ that contains the origin. [More precisely, for each neighborhood N of ∂B_n in the complexification of $\mathbf{C}^n$ there is an $\eta > 0$ so that if H has a holomorphic extension to N that is bounded by η, then $t_0 > 1$.] For such an H let $E(H)$ be the map from $\widehat{B}_n$ to $\mathcal{C}$ obtained by restricting g_1 to $\widehat{B}_n$. From Proposition 8.1 it follows that $E(H) \in \widehat{\mathcal{R}}^\omega$, at least if we shrink the open set of allowable H's as needed.

This exponential mapping is holomorphic on its domain in $\mathcal{V}_0$, and so we can obtain holomorphic maps into $\widehat{\mathcal{R}}^\omega$ from holomorphic maps in $\mathcal{V}_0$. This is pretty good, because $\mathcal{V}_0$ represents the entire formal tangent space to $\widehat{\mathcal{R}}^\omega$ at $\psi_0 = \mathrm{id} : \widehat{B}_n \to \mathcal{C}$, via the correspondence $H \to G \to W$ described above.

Similar remarks hold for any other $\psi_0 \in \widehat{\mathcal{R}}^\omega$, because of Proposition 12.11.

Of course we could also use the generating functions story (from Section 8) instead of exponentiation. This would be a bit messier to describe, but better behaved analytically.

SOME NOTATIONS AND CONVENTIONS (ROUGHLY IN ORDER OF APPEARANCE)

Entity	Explanation, or Location Thereof.
B_n	the unit ball in $\mathbf{C}^n$
ρ	a mapping of B_n into $\mathbf{C}^n$
$d\psi$	the differential of a mapping ψ
bilipschitz	defined in Section 1
λ	an element of $\mathbf{C}$
Δ	the unit disk in $\mathbf{C}$
S^1, S^2	defined in Section 1
D	a domain in $\mathbf{C}^n$ that contains the origin
HCMA	the homogeneous complex Monge-Ampère equation
$\delta\psi(\omega)$	the pull-back of a differential form ω using ψ
completely circled	D is completely circled if $\lambda D \subseteq D$ for all $\lambda \in \overline{\Delta}$
F_0, u_0	$F_0(z) = \|z\|^2, u_0(z) = \frac{1}{2}\log\|z\|$
Green's function	always for the complex Monge-Ampère operator; usually denoted by u, with subscripts, etc.; see (2.4) in Section 2
F	usually $F = e^{2u}$, where u is the relevant Green's function
$i(\cdot)$	interior multiplication (of a vector against a form)
I	a completely circled domain
F_I, u_I, ω_I	Section 3
complex homogeneous of degree 1	Section 3
L_V	the Lie derivative in the direction V
extremal mapping	Section 5
Ψ	Section 5
$\mathcal{C}$, γ, Γ, μ, ν	Section 6
τ	usually as in (6.9), but adjusted to the given situation
Π, Π'	the projections of $\mathcal{C} = \mathbf{C}^n \times \mathbf{C}^n$ onto its first and second sets of co-ordinates
$\widehat{B}_n$, $\hat{\rho}$, $\widehat{D}$	Section 7
$\mathcal{R}^\infty$, $\widehat{\mathcal{R}}^\infty$, $\mathcal{R}^\omega$, $\widehat{\mathcal{R}}^\omega$	spaces of Riemann mappings and their liftings to $\mathcal{C}$; Section 9
$\mathcal{D}^\infty$, $\widehat{\mathcal{D}}^\infty$, $\mathcal{D}^\omega$, $\widehat{\mathcal{D}}^\omega$	spaces of domains and their liftings in $\mathcal{C}$; Section 9
$\mathcal{G}^\infty$, $\widehat{\mathcal{G}}^\infty$, etc.	Section 9
C_a^∞, $\widehat{C}_a^\infty$, etc.	Section 10

$\mathcal{O}^\infty$, $\widehat{\mathcal{O}}^\infty$, etc.	Section 10
$\mathcal{S}^\infty$, $\widehat{\mathcal{S}}^\infty$, etc.	Section 11
S, $\widehat{S}$	Section 11
δ_λ	Section 11
K, R	Section 11
$\mathcal{I}^\infty$, $\widehat{\mathcal{I}}^\infty$, etc.	Section 11
$\mathcal{D}^\infty(I)$, $\mathcal{R}^\infty(\sigma)$, etc.	Section 11
special mappings	Section 11
$\mathcal{H}$	Section 12
$\mathcal{D}^\infty_{co}$	the space of smooth, strongly convex domains in $\mathbf{C}^n$ that contain the origin
$g_D(\cdot,\cdot)$	the Riemannian metric on the space of domains in $\mathbf{C}^n$; see (14.1), (15.17)
u^0, F^0	new name for u_0, F_0; Section 15
Ψ^0	Section 15
T^1, T^2, T^l, $\mathcal{T}^e$	Section 15
$\mathcal{T}$, P	Section 15
$\mathcal{T}_0$, $\mathcal{T}_1$	Section 16
$\mathcal{R}^\infty_{co}$	elements of $\mathcal{R}^\infty$ with image in $\mathcal{D}^\infty_{co}$
$\mathcal{T}_a$, $\mathcal{T}_{a1}$	Section 19
$G_\rho(\cdot,\cdot)$	Section 19
$\widehat{G}_\psi(\cdot,\cdot)$	Section 20

REFERENCES

[A] V.I. Arnold, Mathematical Methods of Classical Mechanics, Graduate Texts in Math., Volume 60, Springer-Verlag, 1978.

[BFG] M. Beals, C. Fefferman, and R. Grossman, *Strictly pseudoconvex domains in* $\mathbf{C}^n$, Bull. Amer. Math. Soc. **8** (1983), 125–322.

[BK] E. Bedford and M. Kalka, *Foliations and complex Monge-Ampère equations*, Communications on Pure and Applied Math. **30** (1977), 543–571.

[C²RSW] R. Coifman, M. Cwikel, R. Rochberg, Y. Sagher, and G. Weiss, *The complex method for interpolation of operators acting on families of Banach spaces*, in Euclidean Harmonic Analysis, edited by J. Benedetto, Lecture Notes in Math. **779** (1980), 123–153, Springer Verlag; *A theory of complex interpolation for families of Banach spaces*, Advances in Math. **33** (1982), 203–229.

[CS] R. Coifman and S. Semmes, *Interpolation of Banach spaces, Perron processes, and Yang-Mills*, to appear, American J. Math..

[D] J.P. Demailly, *Mesures de Monge-Ampère et mesures pluriharmoniques*, Math. Z. **194** (1987), 519–564.

[EM] D. Ebin and J. Marsden, *Groups of diffeomorphisms and the motion of an incompressible fluid*, Ann. of Math. **92** (1970), 102–163.

[K] M. Klimek, *Extremal plurisubharmonic functions and invariant pseudodistances*, Bull. Soc. Math. France **113** (1985), 231–240.

[Ko] S. Kobayashi, Hyperbolic Manifolds and Holomorphic Mappings, Marcel Dekker, 1970.

[L1] L. Lempert, *La métrique de Kobayashi et la représentation des domain sur la boule*, Bull. Soc. Math. France **109** (1981), 427–474.

[L2] ———, *Solving the degenerate complex Monge-Ampère equation with one concentrated singularity*, Math. Ann. **236** (1983), 515–532.

[L3] ———, *Holomorphic invariants, normal forms, and the moduli space of convex domains*, Ann. of Math. **128** (1988), 43–78.

[L4] ———, *Symmetries and other transformations of the complex Monge-Ampère equation*, Duke Math. J. **52** (1985), 869–885.

[M] J. Moser, *On the volume elements on a manifold*, Trans. Amer. Math. Soc. **120** (1965), 289–294.

[R] R. Rochberg, *Interpolation of Banach spaces and negatively curved vector bundles*, Pac. J. Math. **110** (1984), 355–376.

[S] S. Semmes, *Complex Monge-Ampère and symplectic manifolds*, to appear, Amer. J. Math.

[Sl] Z. Slodkowski, *Polynomial hulls with convex sections and interpolation*, Proc. Amer. Math. Soc. **96** (1986), 255–260.

[Wa] F. Warner, Foundations of Differentiable Manifolds and Lie Groups, Scott, Foresman, and Co., 1971.

[We] A. Weinstein, Lectures on Symplectic Manifolds, CBMS Regional conference series in Math., Vol. 29, 1977.

[Wl] R.O. Wells, Differential Analysis on Complex Manifolds, Graduate Texts in Math., Volume 65, Springer-Verlag, 1980.

Stephen Semmes
Rice University
Post Office Box 1892
Houston, Texas 77251

Editorial Information

To be published in the *Memoirs*, a paper must be correct, new, nontrivial, and significant. Further, it must be well written and of interest to a substantial number of mathematicians. Piecemeal results, such as an inconclusive step toward an unproved major theorem or a minor variation on a known result, are in general not acceptable for publication. *Transactions* Editors shall solicit and encourage publication of worthy papers. Papers appearing in *Memoirs* are generally longer than those appearing in *Transactions* with which it shares an editorial committee.

As of May 1, 1992, the backlog for this journal was approximately 8 volumes. This estimate is the result of dividing the number of manuscripts for this journal in the Providence office that have not yet gone to the printer on the above date by the average number of monographs per volume over the previous twelve months. (There are 6 volumes per year, each containing about 3 or 4 numbers.)

A Copyright Transfer Agreement is required before a paper will be published in this journal. By submitting a paper to this journal, authors certify that the manuscript has not been submitted to nor is it under consideration for publication by another journal, conference proceedings, or similar publication.

Information for Authors

Memoirs are printed by photo-offset from camera copy fully prepared by the author. This means that the finished book will look exactly like the copy submitted.

The paper must contain a *descriptive title* and an *abstract* that summarizes the article in language suitable for workers in the general field (algebra, analysis, etc.). The *descriptive title* should be short, but informative; useless or vague phrases such as "some remarks about" or "concerning" should be avoided. The *abstract* should be at least one complete sentence, and at most 300 words. Included with the footnotes to the paper, there should be the 1991 *Mathematics Subject Classification* representing the primary and secondary subjects of the article. This may be followed by a list of *key words and phrases* describing the subject matter of the article and taken from it. A list of the numbers may be found in the annual index of *Mathematical Reviews*, published with the December issue starting in 1990, as well as from the electronic service e-MATH [**telnet e-MATH.ams.com** (or **telnet 130.44.1.100**). Login and password are **e-math**]. For journal abbreviations used in bibliographies, see the list of serials in the latest *Mathematical Reviews* annual index. When the manuscript is submitted, authors should supply the editor with electronic addresses if available. These will be printed after the postal address at the end of each article.

Electronically-prepared manuscripts. The AMS encourages submission of electronically-prepared manuscripts in $\mathcal{A}\mathcal{M}\mathcal{S}$-TEX or $\mathcal{A}\mathcal{M}\mathcal{S}$-LATEX. To this end, the Society has prepared "preprint" style files, specifically the amsppt style of $\mathcal{A}\mathcal{M}\mathcal{S}$-TEX and the amsart style of $\mathcal{A}\mathcal{M}\mathcal{S}$-LATEX, which will simplify the work of authors and of the production staff. Those authors who make use of these style files from the beginning of the writing process will further reduce their own effort.

Guidelines for Preparing Electronic Manuscripts provide additional assistance and are available for use with either $\mathcal{A}_\mathcal{M}\mathcal{S}$-TEX or $\mathcal{A}_\mathcal{M}\mathcal{S}$-LATEX. Authors with FTP access may obtain these *Guidelines* from the Society's Internet node e-MATH.ams.com (130.44.1.100). For those without FTP access they can be obtained free of charge from the e-mail address guide-elec@math.ams.com (Internet) or from the Publications Department, P. O. Box 6248, Providence, RI 02940-6248. When requesting *Guidelines* please specify which version you want.

Electronic manuscripts should be sent to the Providence office only after the paper has been accepted for publication. Please send electronically prepared manuscript files via e-mail to pub-submit@math.ams.com (Internet) or on diskettes to the Publications Department address listed above. When submitting electronic manuscripts please be sure to include a message indicating in which publication the paper has been accepted.

For papers not prepared electronically, model paper may be obtained free of charge from the Editorial Department at the address below.

Two copies of the paper should be sent directly to the appropriate Editor and the author should keep one copy. At that time authors should indicate if the paper has been prepared using $\mathcal{A}_\mathcal{M}\mathcal{S}$-TEX or $\mathcal{A}_\mathcal{M}\mathcal{S}$-LATEX. The *Guide for Authors of Memoirs* gives detailed information on preparing papers for *Memoirs* and may be obtained free of charge from AMS, Editorial Department, P. O. Box 6248, Providence, RI 02940-6248. The *Manual for Authors of Mathematical Papers* should be consulted for symbols and style conventions. The *Manual* may be obtained free of charge from the e-mail address cust-serv@math.ams.com or from the Customer Services Department, at the address above.

Any inquiries concerning a paper that has been accepted for publication should be sent directly to the Editorial Department, American Mathematical Society, P. O. Box 6248, Providence, RI 02940-6248.

Editors

This journal is designed particularly for long research papers (and groups of cognate papers) in pure and applied mathematics. Papers intended for publication in the *Memoirs* should be addressed to one of the following editors:

Recent Titles in This Series

(*Continued from the front of this publication*)

439 **Brian Parshall and Jian-pan Wang,** Quantum linear groups, 1991
438 **Angelo Felice Lopez,** Noether-Lefschetz theory and the Picard group of projective surfaces, 1991
437 **Dennis A. Hejhal,** Regular b-groups, degenerating Riemann surfaces, and spectral theory, 1990
436 **J. E. Marsden, R. Montgomery, and T. Ratiu,** Reduction, symmetry, and phase mechanics, 1990
435 **Aloys Krieg,** Hecke algebras, 1990
434 **François Treves,** Homotopy formulas in the tangential Cauchy-Riemann complex, 1990
433 **Boris Youssin,** Newton polyhedra without coordinates Newton polyhedra of ideals, 1990
432 **M. W. Liebeck, C. E. Praeger, and J. Saxl,** The maximal factorizations of the finite simple groups and their automorphism groups, 1990
431 **Thomas G. Goodwillie,** A multiple disjunction lemma for smooth concordance embeddings, 1990
430 **G. M. Benkart, D. J. Britten, and F. W. Lemire,** Stability in modules for classical Lie algebras: A constructive approach, 1990
429 **Etsuko Bannai,** Positive definite unimodular lattices with trivial automorphism groups, 1990
428 **Loren N. Argabright and Jesús Gil de Lamadrid,** Almost periodic measures, 1990
427 **Tim D. Cochran,** Derivatives of links: Milnor's concordance invariants and Massey's products, 1990
426 **Jan Mycielski, Pavel Pudlák, and Alan S. Stern,** A lattice of chapters of mathematics: Interpretations between theorems, 1990
425 **Wiesław Pawłucki,** Points de Nash des ensembles sous-analytiques, 1990
424 **Yasuo Watatani,** Index for C^*-subalgebras, 1990
423 **Shek-Tung Wong,** The meromorphic continuation and functional equations of cuspidal Eisenstein series for maximal cuspidal subgroups, 1990
422 **A. Hinkkanen,** The structure of certain quasisymmetric groups, 1990
421 **H. W. Broer, G. B. Huitema, F. Takens, and B. L. J. Braaksma,** Unfoldings and bifurcations of quasi-periodic tori, 1990
420 **Tom Lindstrøm,** Brownian motion on nested fractals, 1990
419 **Ronald S. Irving,** A filtered category $\mathscr{O}_S$ and applications, 1989
418 **Hiroaki Hijikata, Arnold K. Pizer, and Thomas R. Shemanske,** The basis problem for modular forms on $\Gamma_0(N)$, 1989
417 **John B. Conway, Domingo A. Herrero, and Bernard B. Morrel,** Completing the Riesz-Dunford functional calculus, 1989
416 **Jacek Nikiel,** Topologies on pseudo-trees and applications, 1989
415 **Amnon Neeman,** Ueda theory: Theorems and problems, 1989
414 **Nikolaus Vonessen,** Actions of linearly reductive groups on affine PI-algebras, 1989
413 **Francesco Brenti,** Unimodal log-concave and polya frequency sequences in combinatorics, 1989
412 **T. Levasseur and J. T. Stafford,** Rings of differential operators on classical rings of invariants, 1989
411 **Matthew G. Brin and T. L. Thickstun,** 3-manifolds which are end 1-movable, 1989
410 **P. L. Robinson and J. H. Rawnsley,** The metaplectic representation, Mp^c structures and geometric quantization, 1989

(See the AMS catalogue for earlier titles)